T0326398

Small Modular Reactors: Nuclear Power Fad or Future?

Related titles

Handbook of Small Modular Nuclear Reactors
(ISBN 978-0-85709-853-5)

**Woodhead Publishing Series in Energy:
Number 90**

Small Modular Reactors

Nuclear Power Fad or Future?

Daniel T. Ingersoll

ELSEVIER

AMSTERDAM • BOSTON • CAMBRIDGE • HEIDELBERG
LONDON • NEW YORK • OXFORD • PARIS • SAN DIEGO
SAN FRANCISCO • SINGAPORE • SYDNEY • TOKYO
Woodhead Publishing is an imprint of Elsevier

WP

WOODHEAD
PUBLISHING

Woodhead Publishing is an imprint of Elsevier
80 High Street, Sawston, Cambridge, CB22 3HJ, UK
225 Wyman Street, Waltham, MA 02451, USA
Langford Lane, Kidlington, OX5 1GB, UK

Notices
Knowledge and best practice in this field are constantly changing. As new research and experience broaden our understanding, changes in research methods, professional practices, or medical treatment may become necessary.

Practitioners and researchers must always rely on their own experience and knowledge in evaluating and using any information, methods, compounds, or experiments described herein. In using such information or methods they should be mindful of their own safety and the safety of others, including parties for whom they have a professional responsibility.

To the fullest extent of the law, neither the Publisher nor the authors, contributors, or editors, assume any liability for any injury and/or damage to persons or property as a matter of products liability, negligence or otherwise, or from any use or operation of any methods, products, instructions, or ideas contained in the material herein.

ISBN: 978-0-08-100252-0 (print)
ISBN: 978-0-08-100268-1 (online)

British Library Cataloguing-in-Publication Data
A catalogue record for this book is available from the British Library

Library of Congress Cataloging-in-Publication Data
A catalog record for this book is available from the Library of Congress

For information on all Woodhead Publishing publications
visit our website at http://store.elsevier.com/

Working together
to grow libraries in
developing countries

www.elsevier.com • www.bookaid.org

To Katie,
my loving wife and
daily reminder of what is most important in life.

Contents

Woodhead Publishing Series in Energy

Foreword

When I was asked to write a foreword for a book on small modular nuclear reactors (SMRs), I thought: "This book could not come at a better time or from a more knowledgeable person!"

Energy production is surging globally, even though it has flattened in the US and Europe. Much of the world now admits that we need to address global environmental changes caused by generating over 17 trillion kWh of electricity each year and by burning 90 million barrels of oil each day. These numbers will double in the next 25 years.

This doubling of energy consumption is actually necessary and ethical.

A human requires 3000 kWh per year to achieve what we consider to be a good life. The enormous effects of the industrial revolution came about as a result of the availability of energy from sources other than muscle: first coal, then oil, gas, hydroelectric, nuclear, and now renewables. Suddenly, anyone could obtain 3000 kWh per year without subjugating or enslaving other humans. This led to something previously unknown in history: the middle class. This rapidly emergent group also demanded expanded social and civil rights.

One can see the same effects going on in the developing world today. Modern China was born in the early 1990s with the planned installation of almost 600 coal-fired power plants, along with large hydroelectric, which led directly to China's present middle class population of 500 million people. This process of electrification of human society is continuing, with a goal of 30 trillion kWh annually worldwide—the minimum amount of energy needed to eradicate global poverty.

The question is how to provide this much energy without destroying the environment. Coal requires the least infrastructure to emplace. Thus, the energy-hungry developing world is keeping coal in the position as the fastest growing energy source globally. Natural gas is second. Nuclear, hydro, and renewables bring up the rear. For many reasons, we need to reverse this trend.

It is the extreme energy density of nuclear power that is at the heart of our global energy solution and nuclear power's low environmental impacts. The toxic waste generated by a 1000-MW coal plant is 10 million times as voluminous as the waste generated from a 1000-MW nuclear plant. The carbon emitted from coal plants is almost 100 times greater than that of nuclear plants for the same energy produced. A 1000-MW nuclear reactor on a one-square-mile site will produce the same amount of energy over its lifetime as 10,000 1-MW wind turbines on 1500 square miles.

Because nuclear reactors run for so many decades, the actual lifetime costs of nuclear energy are the second lowest of all sources (short-term finance and energy market issues aside), second only to hydroelectric.

Because of these properties, 72 new nuclear reactors are under construction around the world, and 150 more are firmly planned. China is hoping to replace 300 of their coal plants with nuclear plants by the mid-century. India is planning 100 new nuclear reactors over the next 30 years.

Those who know the value of energy know that nuclear and hydroelectric power are the only sources that have competed with fossil fuels in any significant sense globally. Hydro is still growing in the developing world but is fast approaching its geographic limits and is vulnerable to droughts. Nuclear is nowhere near its limit and is mostly immune to climate and weather changes. With new nuclear reactor designs, both large and small, there is sufficient nuclear fuel for thousands of years of energy, even at 30 trillion kWh per year.

Our experience with nuclear energy over the last 50 years has proven that nuclear is the safest and most efficient of all energy sources, from both the human health and environmental perspectives. To produce a trillion kWh of electricity, nuclear takes less land, uses less steel and concrete, harms fewer people, and has fewer emissions than any other energy source, including wind and solar.

But we are at the point where expanding the existing nuclear capacity will require new nuclear plant designs that are more flexible in size, cost, location, applications, and operation and are safer than even the extremely safe reactors we have now. They must also be able to backup, or load-follow, the most intermittent of renewable sources.

Enter small modular reactors.

The new generation of reactors embodied in SMRs incorporates all of the experience and technological advances in the last 50 years and is poised to enter the market in a big way. But confusion abounds as to their place in our energy tool box.

Dr Ingersoll knows the scientific, social, and economic challenges of significantly increasing nuclear power in the world and the role of SMRs in the future. He carefully addresses each issue in this book.

We are at energy crossroads in the history of humanity that will determine what will be left of our planet in the twenty-second century. Understanding SMRs is a necessary step to make the correct decisions that we need to retain a beautiful world and provide all humans with the opportunity to enjoy it.

Understanding SMRs is what this book is all about.

James Conca
Science Contributor to Forbes

Preface

I find writing neither easy nor pleasurable. Scientists and engineers are not generally known for their communication skills, and I am no exception. But I have had a steadily growing interest in documenting my excitement for small modular reactors (SMRs) for more than 10 years. I have written and coauthored a number of papers on selected SMR topics and coedited with Mario Carelli a collective handbook on SMRs authored by a broad spectrum of experts. But these projects only served to feed a growing passion to capture my complete thoughts on the subject, cover-to-cover. What pushed me into action was a series of events and announcements in the early 2014 that seemed to cast undue doubt on the credibility of SMRs and their ability to succeed in the marketplace. The purpose of this book is to refute that doubt and to present the basis for my fervent belief that SMRs will in fact become an important part of our energy future.

So, this is an advocacy book for SMRs. It is not generally intended for the SMR practitioner, although it may offer a few new nuggets of interest to those already working in the field. Mostly, it is intended for those who know very little about SMRs or who are skeptical about their viability and are interested in learning more. To improve readability, I have attempted to present the information in a more personal, engaging manner than most technical handbooks and journal papers. I also make frequent use of analogies. Like most analogies, none are perfect, and some may seem a bit contrived. But I find analogies very helpful when trying to learn a new topic and hope that it helps your understanding as well.

In this book I share many of my own discoveries and learnings during the past several years. I have drawn as much as possible from my own direct experience, which unavoidably introduces a personal bias. For example, most of the book's content is focused on SMR development in the US. Also, there is a considerable focus on water-cooled SMR designs, partly because I have been most engaged with that class of SMRs and partly because I believe that they will be the first to reach the marketplace. I provide a brief summary of my significant SMR involvements below as a contextual backdrop for the book.

My fascination with SMRs has been a series of progressively reinforcing encounters—each one measurably nudging me toward a new career path. It began in the 1980s when, as a radiation-shielding researcher at Oak Ridge National Laboratory, I was tasked with analyzing the shielding requirements of the Power Reactor Inherently Safe Module (PRISM) design being developed by General Electric. PRISM was a significant departure from traditional commercial reactors not only because it was cooled with sodium rather than water, but also because it took an entirely different approach to plant design. In particular, a large output plant was comprised of nine small modules. This novel approach to delivering nuclear power immediately intrigued me, primarily due to the design simplifications and plant flexibilities enabled by this approach.

In the fall of 2000, I attended a colloquium at the University of Tennessee presented by Mario Carelli, a chief scientist at Westinghouse. Carelli described the newly formed international consortium that was developing the International Reactor Innovative and Secure (IRIS) design. IRIS was a small, integral light-water reactor that focused on achieving an exceptionally high level of safety—"safety by design," as Carelli liked to call it. After appropriate due diligence, Oak Ridge National Laboratory elected to join the IRIS consortium, and I served as the laboratory's principal contributor. I remained actively engaged with IRIS for several years until a new opportunity surfaced: the Global Nuclear Energy Partnership.

The US Department of Energy unveiled the Global Nuclear Energy Partnership program in 2006. Although the majority of the program was focused on developing new technologies for separating and recycling nuclear waste, there was a minor program element to explore smaller-sized reactors for international deployment. The Grid-Appropriate Reactors program catalyzed significant interest in SMRs both within the US and internationally. My position as a technical leader for the Grid-Appropriate Reactor program significantly shaped my view of the global energy landscape and the vital role that SMRs could play. The program was short-lived, but I was able to continue working with the DOE to develop the follow-on, domestically focused Small Modular Reactor program. This engagement provided a new insight into the spectrum of SMRs being designed in the US and a surprising interest in SMRs by domestic utilities.

A few years later, I chose to leave the Oak Ridge National Laboratory and join NuScale Power, a startup company dedicated to commercializing a novel and innovative SMR. This event irreversibly changed my career path from being involved in advanced reactor research to being committed to the development and deployment of SMRs, specifically the NuScale SMR. Although I was employed by NuScale Power during the writing of this book, it was prepared entirely on my own time and at my own expense. The opinions given in it are exclusively my own and are not intended to reflect the opinions of NuScale Power. I have tried to not let my allegiance to NuScale Power unfairly bias my writing, although I frequently use the NuScale design as an illustrative example due to my familiarity with that particular SMR design. On the other hand, I freely acknowledge that working at NuScale Power was a deliberate choice based on having studied SMRs of all technologies and designs for many years. I believe that the NuScale design best embodies the desirable features of SMRs and will best demonstrate their many anticipated benefits. I am not alone in my excitement for the design; the halls at NuScale Power are filled with new colleagues who have also made tough personal decisions to dedicate their time and talent to that effort. I am both honored and excited to be a part of the NuScale team.

Reflections of these past engagements with SMRs, domestic and international, appear throughout the book, which is divided into three major parts. In Part One, I describe the global energy landscape and offer perspectives on why nuclear energy will be an important contributor to our global energy future. Within this discussion, I touch on the strengths and weaknesses of several promising energy options and introduce notions of how SMRs can complement those options. I follow the introduction with a historical review of SMRs during the first 50 years of the nuclear power industry. Smaller-sized reactors played a major role in the initiation of the industry, and many SMR designs were developed throughout those 50 years, although they never quite made it to the marketplace. In a separate chapter, I review the recent history of SMR activities during the past 15 years—the portion of their history that I have experienced firsthand.

Part Two of the book addresses what I believe to be the three primary tenets of SMRs: enhanced safety, improved affordability, and expanded flexibility. These chapters speak to the root of my enthusiasm for SMRs. It should be apparent in the discussions that not all SMRs are created equal, that is, different designs capitalize on different features and offer different benefits. But in general, the smaller unit size of an SMR provides better opportunities to add additional accident resilience into the design and allows for greater flexibility in how nuclear power is used for diverse applications. Some SMRs also offer design simplifications and small, incremental capacity growth features that can significantly improve nuclear power affordability.

The third and final set of chapters discusses the issues associated with the deployment of SMRs, including customer interest and remaining challenges. While the level and breadth of customer interest in SMRs continue to expand, there are many remaining hurdles to be cleared before getting them into the commercial market, including technical, institutional, and social challenges. Within the discussion of remaining challenges, I highlight several potential opportunities to meet those challenges in new and innovative ways. Finally, I also revisit some of the key arguments that support SMRs as either a short-term fad or a vital part of nuclear power's future. I close the book on the same critical issue that I start with: the undeniable and ever-increasing need for energy. I hope this book helps you to conclude, as do I, that nuclear power will play an important role in meeting that need and that SMRs will be a pivotal and enduring part of nuclear power's future.

Acknowledgments

My pursuit of SMRs has allowed me to meet and work with many talented and inspiring colleagues. I have mentioned several of them within the text of the book and offer special acknowledgment and gratitude to a few of them here.

I would first like to acknowledge Mario Carelli, the dynamic leader of the IRIS international consortium. I learned a tremendous amount of reactor engineering from Mario and the IRIS team. I also wish to acknowledge the group from the Politecnico di Milano led by Marco Ricotti. The POLIMI group contributed to nearly every facet of the IRIS design and continues to produce insightful studies on various aspects of SMRs.

I would like to acknowledge the effective leadership and persistence of Dick Black at the US Department of Energy (DOE) in securing federal funding for the SMR program. I am also grateful to Dick for providing a peer review of my manuscript. Other notable colleagues at the DOE include Rob Price, who managed the Grid-Appropriate Reactors program, and Tim Beville, who manages the SMR Licensing Technical Support program. I also wish to acknowledge Sherrell Greene for his generous support at Oak Ridge National Laboratory that allowed me to engage with the SMR program despite extended delays in federal funding.

Internationally, two colleagues stand out for their contributions to encouraging broad awareness of smaller-sized reactors: Vladimir Kuznetsov and Hadid Subki. They conducted several programs at the International Atomic Energy Agency to introduce SMRs to embarking countries and support a sustained dialog among the growing SMR community.

I especially want to express my gratitude to Jose Reyes, cofounder of NuScale Power. He enthusiastically welcomed me to NuScale and has been a pleasure to work with. I continue to be inspired by his technical leadership, professional dedication, and personal integrity.

I wish to thank Woodhead Publishing for agreeing to publish this book and particularly Sarah Hughes for her encouragement and constructive feedback during the manuscript development. I also thank Alex White for his help with the final production process.

Most importantly, I thank my beautiful wife, Katie, for her loving patience and encouragement throughout this project and her review and critique of my initial drafts. I am eternally grateful for her devoted support of my career pursuits despite significant disruptions to our family. I also thank my three fantastic children, Ben, Tina, and Lori, for enduring my many physics lectures and for blessing Katie and me with six delightful grandkids. Finally, in loving memory, I thank my mother and father for their sacrifices to give me valuable opportunities early in my life.

Part One

Setting the stage

Energy, nuclear power, and small modular reactors

1

For those who are new to the topic of small modular reactors, or SMRs, as we like to call them, they are simply a different way to package nuclear energy to produce heat or electricity for commercial energy markets. They are neither new to the nuclear industry nor represent a whole new technology. In an analogy to the automotive industry, they represent a Smart Car in an industry that is dominated by Hummers. They are intended to offer an energy option to those customers for whom large nuclear plants are not a viable choice. Their design features are driven by the needs of those new customers, who generally require lower sticker prices, higher levels of plant resilience, and greater flexibility of siting and use. SMRs have gained considerable attention and have generated both a lot of excitement and a lot of confusion in the nuclear industry. The intent of this book is to offer my personal perspectives, derived from 15 years of immersion in the topic, on the basis for the excitement and clarification of the confusion. Most importantly, I hope to provide sufficient evidence and insight to answer the question at hand: Are SMRs just a market fad, or will they be an enduring part of the future of nuclear power?

1.1 Fad or future?

I have been engaged in research and development of advanced nuclear power for most of my career. During the past 15 years, I have focused almost entirely on SMRs. I have been so absorbed in the subject and so certain of their merit that it never occurred to me that SMRs might not ever reach the marketplace. I worked with the US Department of Energy from 2009 to 2012 to build a new program to encourage research, development, licensing, and deployment of SMRs, and we saw rapidly expanding interest for them from the nuclear community. We also saw unprecedented political support from the White House, from the Senate and House of Representatives, and from both the Republican and Democratic parties. The industry press and even the popular press were filled with optimistic articles on the exciting promises of SMRs. Sessions at industry conferences emerged that focused on SMRs, and entire conferences were dedicated to the subject. Even the tenacious antinuclear community levied only weak counterarguments centered mostly on the yet-to-be-demonstrated benefits of SMRs.

At the beginning of 2014, two prominent US SMR vendors made public declarations that they were significantly reducing their efforts to develop and license new SMR designs, ostensibly based on financial considerations and market uncertainties. Very quickly, articles began to surface stating that SMRs were just a fad and that they would soon be destined for deployment to history books. This notion was completely foreign to me and spurred me to ponder: Could SMRs be just a fad? Also, given where

Small Modular Reactors. http://dx.doi.org/10.1016/B978-0-08-100252-0.00001-X

we are in their development, can we reliably predict if they are likely to fade away or become a permanent part of our energy future?

In addressing these questions, I find it useful to inspect previous fads and to look for common trends. The toy industry is a target-rich environment for studying fads. One example, Pet Rocks, barely survived a single holiday season in 1975. To be honest, I was immediately skeptical about the product's longevity—after all, why would anyone pay several dollars for a common river rock? Cabbage Patch dolls survived a couple of years during the 1980s, but they too faded rather quickly into product history. Similarly, the initial popularity of Teenage Mutant Ninja Turtles peaked and then faded a few years later, although it appears that the hard-shelled warriors are making a comeback.

In contrast, other new products, although initially viewed as short-lived fads, have ended up changing the future of their market or even creating entirely new markets. Example: the iPad tablet. When introduced in 2010, the Apple iPad was quickly labeled as a fad, and its sales were attributed to the tenacity of Apple fanatics, who routinely stand in line for days to be one of the first purchasers of Apple's latest gizmo. I happen to work with some of these fanatics and overheard one of them exclaim: "I didn't even know that I needed it until I saw one." A few short years later, tablet computers dominated the consumer electronics market and have had a profound impact on shaping the features of cell phones, laptop computers, and mainstream operating systems. Our seven-year-old grandson already has his own tablet and frequently instructs me on its use.

Why the difference, and how do we know what will be fad or future? From the simple examples above, I have to conclude that uniquely satisfying a consumer need is a key ingredient. The example of the iPad provides the most insight—the product addressed a consumer need that was not even fully appreciated until it was met. In the chapters that follow, I will attempt to walk through the "fad or future" question for the case of SMRs with an emphasis on highlighting their characteristics that address specific customer needs. I freely admit from the outset that I am a huge believer in nuclear power in general and SMRs in particular. So rather than try to disguise this bias, I will offer the basis for my enthusiasm. I will also appropriately temper my enthusiasm with a discussion of their limitations as well as the many hurdles they face to realize their promise.

The challenge in predicting the future of new products in the nuclear field is that advancements move at a glacial pace and can take decades to implement. Actually, even glaciers may be outpacing the nuclear industry now. On a recent Alaskan cruise, my wife and I were awed and alarmed by the fact that the glaciers at the ends of those majestic fjords have receded significant distances—miles in some cases—in just the past few years. I am not sure the nuclear industry can boast such a discernable rate of change. I would argue that it is this slow pace of progress that most contributes to the notion by some that SMRs are a fad. This happens because people are inherently impatient, an increasing trait in our world of instantaneous communications. Some people have been too quick to claim SMRs as a success, which creates a feeling of disillusionment as progress edges forward at a snail's pace. Other people are too quick to predict their failure, which is sometimes a result of using the wrong metrics to judge success and failure. Referring back to the example of the iPad, there were those who discredited the iPad when it was first released because it was much bulkier than a cell phone yet was too limited in processor power to be a computer. Basically, the iPad was being judged by the same metrics to evaluate cell phones and desktop computers. The iPad has succeeded because it was not intended to be either a cell phone or a desktop computer—it offered something entirely new. This is a common trait of critics, that is, to focus on what a product cannot do, or does not do well, rather than to see the potential for what it can do. Unfortunately, the slow rate of development in the nuclear industry, which is measured in decades, allows for a much more protracted attack by critics on the value proposition of new developments such as SMRs.

Fad or future? It will take several more years to know for sure. Based on the success of the iPad, a more telling question might be: Do SMRs satisfy an important unmet need? A hint at the answer may be hidden in the global energy landscape, including trends in nuclear power as a part of the global energy mix. I start my story with a look at the need for energy.

1.2 The importance of energy

The reason for considering nuclear power, or any energy source, is quite simple: the availability of abundant, affordable energy is directly linked with quality of life. Multiple studies have shown a tight correlation between energy consumption and quality of life. A leading study conducted by Alan Pasternak at Lawrence Livermore National Laboratory[1] demonstrated this correlation in a compelling way using the human development index (HDI), which is a complex, multifaceted metric maintained by the United Nations to annually evaluate the quality of life for every country in the world. The HDI is based on several national factors, including such indicators as life expectancy, nutrition, income, years of education, and access to clean water resources.[2] Pasternak plotted the HDI value for 60 countries as a function of the country's annual per capita electricity consumption. What he observed was a striking correlation between the two factors with an apparent threshold of 4000 MWh electricity consumption for HDI values greater than 0.9; nearly all countries with a relatively high quality of life consume at least 4000 MWh per capita annually.

Pasternak's study was based on 1997 data for HDI and electricity consumption. I have updated his results in Figure 1.1, which presents a similar HDI versus per capita energy consumption for 129 countries based on 2011 data.[2,3] As in Pasternak's original study, the figure shows a clear correlation between a country's HDI and its per capita energy consumption. There is a modest amount of scatter in the data, which is due in part to the impact of climate conditions experienced in the different countries, that is, high-latitude countries tend to consume more electricity per capita than countries with more temperate climates. For instance, Iceland, which was excluded from the figure because it significantly distorts the scale of the graph, uses a sizable 52,400 kWh per capita to achieve an HDI of 0.89 compared to Japan, which uses just 7900 kWh per capita to achieve a similar HDI value. But the trend is unmistakable. It is this importance of energy for improving quality of life that originally enticed me into studying nuclear power nearly 40 years ago, and it is what continues to motivate me today.

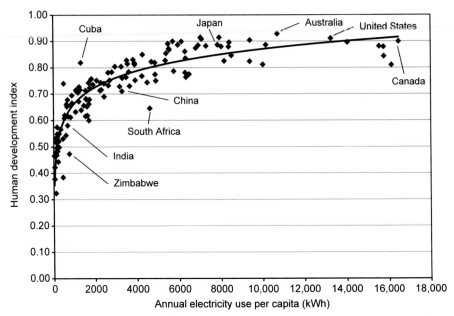

Figure 1.1 Relationship between per capita energy consumption and quality of life (2011 data). The trend line is a simple logarithmic fit to the 129 data points.[2,3]

It is worth noting in Figure 1.1 that China and India, the two most populous countries in the world, are at a low-to-modest HDI level and a correspondingly low rate of energy consumption. Both factors are expected to increase in the coming years as these two countries add new energy capacity at a staggering rate. In fact, China has tripled its energy consumption per capita between Pasternak's 1997 data and the updated 2011 data. India has also seen an increase in energy consumption, although less dramatic than in China.

In the spirit of full disclosure, a correlation does not necessarily imply a causal effect. For instance, the fact that more ice cream is consumed on days that more sunburns occur does not mean that ice cream causes sunburns. In reality, both factors are a consequence of the same shared cause: a hot, sunny day. So the relationship observed by Pasternak does not necessarily mean that more energy consumption creates a higher quality of life or vice versa. However, as I look around me at the features in my home that contribute to my high quality of life, I find it intuitively obvious that the two factors are causally related. Other studies that are more scientific than my simple observations have also concluded a strong causal relationship, especially regarding access to high-quality energy such as electricity and natural gas.[4]

The relationship between energy and quality of life is certain to become even more pronounced as clean water becomes increasingly scarce. A major contributor to the HDI is access to clean water, yet more and more regions and entire countries have significant water shortages or are consuming fresh water faster than it is being resupplied. Countries throughout the world, including the US, are striving to reduce or eliminate their import of energy sources, while also facing increasing shortages of clean, potable water. Scores of countries are considered to be "water stressed," meaning that their availability of fresh water is below 2000 m³ per person per year. Even in countries that have adequate water resources nationally, the geographic distribution of water typically is not uniform and selected regions may suffer from water scarcity. An example of this is the southwestern region of the US, where the annual consumption of water has been exceeding water production for the past several years.[5] Even in the normally precipitation-rich southeastern region of the US, droughts and population growth have created severe water shortages in some locations, resulting in regional tensions. As an example, I live in eastern Tennessee, and of local interest are the repeated attempts by the Georgia state legislature to change the location of the long-standing boundary between Georgia and Tennessee. They claim that the current boundary is the result of a surveyor's mistake in 1818 and incorrectly denies Georgia access to desperately needed water flowing in the Tennessee River, which is just one mile north of the errant border.[6]

As access to clean ground water and surface water sources dwindle, more regions are turning to water desalination as a means to meet clean water demands. According to the Global Water Intelligence organization, approximately 16,000 desalination plants exist worldwide, producing roughly 75 million cubic meters per day.[7] Over 700 new plants were added in 2010–2011, which collectively increased the global capacity by 5.2 million cubic meters per day. This growth is expected to continue and is being driven by continued population growth, rapid industrialization in developing countries, urbanization, and dwindling fresh water sources.

The dilemma: it takes water to produce energy and it takes energy to produce and distribute water. This has spurred considerable interest in what is sometimes referred to as the energy–water nexus.[8] For example, 41% of the US' fresh water withdrawal in 2005 was used for cooling thermoelectric power plants. That same year, an average of 7300 MW of power was used globally to produce 35 million cubic meters per day of clean water. With many tens of countries already facing extreme water scarcity—the same countries that are also expecting to increase economic development and quality of life—the demand and competition for abundant, affordable energy will become fierce.

The world population continues to increase, but the desire to improve quality of life is increasing even faster. The US Energy Information Administration (EIA) projects that global energy consumption will grow by 56% in the next 30 years—just one generation.[9] The EIA further predicts that most of the growth will be in countries whose current economic impacts are barely on the radar screen of the world market. Specifically, they estimate that the energy consumption in developing countries will nearly double, while developed countries will experience a modest 17% increase in energy consumption. The huge energy demand in developing countries will be driven by, or be a consequence of, their pursuit of a better life.

There is a considerable amount of variation in energy policy among countries, due in part to different social motivators. In the case of India, they are anxious to improve their living conditions and have concluded that their low HDI rating can only be addressed through a deliberate program to expand their energy production capabilities. They started with a targeted quality of life for their population, and then using Pasternak's relationship, or something similar, determined how much additional energy per capita would be required to achieve their target HDI.[10] Given their population of over one billion people (and rising), this is a staggering amount of energy—many TW of new generating capacity. As part of their energy growth strategy, they have embarked on an aggressive three-phase program to greatly expand their nuclear energy capacity, which will result in a sustainable, closed fuel cycle based on their substantial domestic reserves of thorium.

China is also in the throes of a rapid expansion of their economy. They are building new power plants of all types as fast as they can collect the raw materials. Currently they have nearly 400 coal plant projects planned, the fastest growing rate of new solar capacity in the world, and 28 nuclear plants under construction.[11] In terms of their nuclear program, they are not only rapidly expanding their fleet of water-cooled reactors, using both imported and domestic designs, but they are also leading the world in the pursuit of advanced reactor technologies. In particular, they are building a sodium fast reactor demonstration plant, two high-temperature gas-cooled reactors, and are embarking on a molten salt test reactor project. Other countries of Southeast Asia, such as Malaysia, Indonesia, and Vietnam, are also expanding their economies rapidly, and most are actively pursuing nuclear power.[12]

Western Europe already has a highly developed economy with a high quality of life and a mature energy infrastructure. Their interests for new energy capacity are primarily to maintain their economies and improve energy security while reducing the use of carbon-emitting energy sources. Some countries, such as Germany, are attempting to achieve these goals with the increased use of wind and solar energy sources. Countries such as France are holding strong to their commitment to nuclear power—a decision that they have sustained since 1974. The United Kingdom, once a major global pioneer in nuclear power, had moved away from this energy option but are now re-embracing nuclear power quickly. At the same time, Eastern European countries are accelerating their economic development as they integrate into the European community and work to be self-sufficient in terms of energy production capacity.

The countries of the Middle East, known for their vast oil resources, are surprisingly turning to nuclear power for future energy development. Although many other countries have become hopelessly dependent on oil imports to meet their energy needs, the Middle Eastern countries are equally dependent on the income from their oil exports. Their realization of this dependence is motivating many of these countries to develop domestic nuclear power programs in order to conserve finite oil resources for oil-thirsty countries like the US. The United Arab Emirates is currently building its first four nuclear power plants and several other countries in that region, led by Saudi Arabia, Jordan, and Egypt, are actively pursuing nuclear power programs.[12] In total, nuclear power is projected to more than double globally in the next 30 years, bolstered by considerations of economic development, energy security, and greenhouse gas (GHG) emissions.[9]

The US generally enjoys a quality of life that is the envy of most of the world. Perhaps because of a success-bred complacency, the US continues to lack a definitive, rational, and sustainable energy policy, despite several major disruptions to the flow of energy imports, domestic energy supplies, and infrastructure. I remain optimistic, though, since there appears to be increased attention regarding the importance of pursuing a diverse portfolio of clean energy options. As Winston Churchill was quoted, "The Americans always get it right, but only after they've exhausted all other possibilities." As evidence of the progress toward a sensible energy strategy, in his State of the Union address in 2011, President Obama asserted the following:

> *By 2035, 80% of America's electricity will come from clean energy sources. Some folks want wind and solar. Others want nuclear, clean coal and natural gas. To meet this goal we will need them all.*[13]

The US is fortunate to have many energy resources and the affluence to manufacture or purchase other clean energy technologies. Many countries do not have this luxury, and their portfolios are quite limited. Early in my career, I foolishly argued to justify why nuclear was so much better than other energy sources. I've come to appreciate, as do most energy planners, that a balanced energy portfolio of multiple options is the right answer. As the President now asserts, we need "all of the above."

To understand the options, especially for reducing GHG emissions, it is important to understand the entire life cycle GHG emission characteristics for the many energy options available to the current energy mix. Figure 1.2 shows the results of an extensive study of life cycle GHG emissions by the World Nuclear Associations based on a compilation of 21 independent studies.[14] The range of emission values for each source is due to the varying assumptions within each of the studies and the diversity of processes used to mine or recover the raw fuel material, manufacture and construct the plant infrastructure, deliver the energy, and dispose of the waste. The data are quite convincing that nonfossil energy sources have the greatest potential for reducing GHG emissions—not a big shocker. Unfortunately, two-thirds (67%) of the electricity in the US comes from fossil-based fuels, as shown in Figure 1.3, which characterizes the

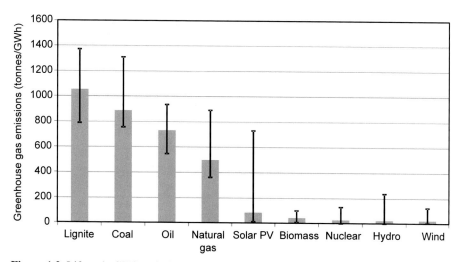

Figure 1.2 Lifecycle GHG emissions for several different energy sources compared on a common basis of electricity production.[14]

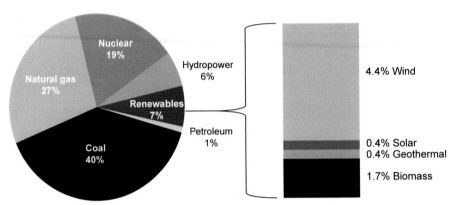

Figure 1.3 Major sources of electricity production in the US in 2014.[15]

mix of energy sources used by the US in 2014 to generate electricity.[15] If we add in transportation and industrial energy uses, the percentage of the total energy production from fossil-based fuels rises to 80%. Houston, we have a problem.

The US is actively working to expand its use of renewable energy sources. We should continue to do this while being mindful of some fundamental challenges with these seemingly "free" energy sources. As skillfully presented by William Tucker in *Terrestrial Energy*, these sources fight a huge handicap of energy density.[16] By Tucker's math, nuclear energy has an intrinsic energy density that is two million times larger than fossil fuels, which in turn have energy densities that are 2–50 times larger than various renewable sources (wind, solar, and biofuels). To compensate, renewables such as wind and solar energy must be collected over vast areas of land. According to Gwyneth Cravens, author of *Power to Save the World*, a wind farm producing 1000 MWe would require 200 square miles, a solar array producing the same amount of electricity would require 50 square miles, and a nuclear plant would require a third of a square mile.[17] In some regions of the country, this may be an acceptable trade-off. The vast sun-parched regions of the Southwest offer a great location to collect the sun's energy using either photovoltaic or concentrated solar technologies, and the broad farmlands of the Midwest can be reasonably shared with wind farms. However, it is harder to imagine this working on Manhattan Island or in downtown Atlanta. In contrast, nuclear energy offers a virtually unlimited supply of abundant, clean energy with a track record of safety that is unsurpassed in the energy industry. The time and conditions are right for the US to join the global bandwagon for the increased use of nuclear power.

1.3 New growth of nuclear power: the nuclear renaissance

Beginning around the year 2000, there seemed to be a resurgence of interest in nuclear power within the US. It is what Alvin Weinberg, an early pioneer in the nuclear industry, predicted as the "second nuclear era."[18] More commonly, this resurgence of interest is referred to as the "nuclear renaissance" by industry enthusiasts. The nuclear renaissance is mostly a US phenomenon. The early start of commercial nuclear power in the US allowed it to build a substantial generating capacity through the 1970s. Unfortunately, this was followed by three decades of no new plant orders and many plant cancellations or closures. The US now appears to be firmly back on the path of expanding its nuclear generating capacity with the completion of a previously abandoned nuclear unit (Watts Bar Unit 2) and the construction of four new units: two units at the Vogtle site in Georgia and two at the V.C. Summer site in South Carolina. Other countries, such as France, China, and India, have continued to grow their nuclear fleet with no apparent pauses. Still other countries, especially those who are seeking to initiate nuclear power programs, are closely watching the US position toward nuclear power and waiting for clear signs that its nuclear renaissance is real.

Given the excellent track record for safety and reliability by the US nuclear industry, nuclear power should be well positioned to become a significant portion of the US clean energy mix. There are several reasons why now is the time for the US to re-embrace nuclear power. France uses nuclear power to produce nearly 80% of its electricity, while the US generates less than 20% of its electricity from nuclear. In the past, the biggest challenge to following the French lead was the US' large reserves of cheap coal, which is why coal combustion has dominated its energy mix. But beginning with the Kyoto Protocol, which was drafted in 1997 and implemented in 2005, there has been growing global concern for the deleterious impacts of GHG emissions, principally carbon, into the atmosphere. Whether you are a staunch believer in global climate change, or someone who appreciates clean air, it just doesn't make sense to belch millions of tons of carbon into the air every day. The US has made on-again, off-again efforts to reduce carbon emissions, but there now appears to be a more sustained and concerted effort to address this critical goal.

During Barak Obama's campaign for presidency in 2008, he promised to reduce US carbon emissions by 80% by 2050 relative to 2005 emissions. He followed that promise in 2009 with Executive Order 13514, which established very rigorous emission reduction goals for federal facilities.[19] To demonstrate the size of this challenge, the US carbon emission in 2005 was roughly 6000 Tg (or 6000 million metric tons).[20] An 80% reduction in that amount leaves us with a target of 1200 Tg. The last time in history that the US emitted that level of carbon into the atmosphere was 1906![21] Truly, the challenge in achieving this essential goal, while not significantly impacting the quality of life, is staggering.

In the power production industry, coal plants are the worst offenders for emission of GHGs. Coal plants in the US account for 75% of CO_2 emissions from the electricity production sector.[20] Even with modern scrubbers, coal plants are very challenged to significantly reduce carbon emissions. "Clean coal" technology, an oxymoron in some people's minds, has many technical hurdles, as does the hope of carbon capture and sequestration. Research in these areas is continuing, and should, but progress is proving to be very challenging, and the more promising approaches appear to be quite expensive. Meanwhile, many utilities across the country are beginning to program the closure of their coal plants, especially the older, less efficient plants. Some of the big coal-producing states are concerned about this trend and are already working to develop technologies that can convert coal into higher value (lower emission) products.

The lower portion of Figure 1.4 provides a scatter plot of all US coal plants operating in 2008 as a function of their initial date of operation and nameplate capacity.[22] The upper portion of the figure is a cumulative capacity curve for those coal plants. If US utilities were to close all coal plants that are more than 50 years old, they would need to replace roughly 75 GWe of capacity. This number jumps an additional 100 GWe if plants older than 40 years are closed. Replacing this amount of base-load electricity is a huge challenge.

One element of a solution to the GHG emission challenge is to increase the use of renewable energy sources, which include hydroelectric, wind, solar, geothermal, and biofuels. The US has increased its use of nonhydro renewables by roughly a factor of three in the past 10 years, although they still represent only 6% of its total capacity, two-thirds of which is from wind.[23] Despite the caveat that I mentioned earlier regarding energy density, I think the expanded use of renewables is a positive trend *where it makes sense to do so*. As a supporting example, I was driving through the Columbia River Gorge between Washington and Oregon on an October day in 2012 when it was so windy that I had to drive with both hands firmly on the wheel as I passed by thousands of whirling wind turbines. Coincidentally, a power generation record was set that same day by those countless wind turbines, which are plugged into the Bonneville Power Authority's system. It was the first time in history that the wind farm output

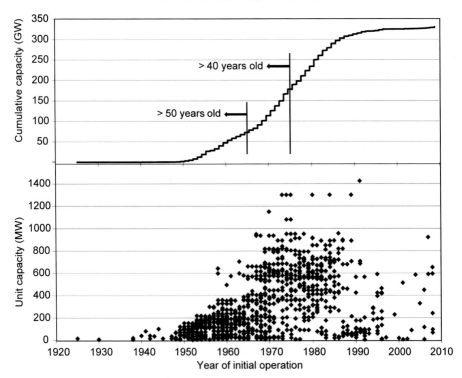

Figure 1.4 Output capacity of all US coal plants (lower graph) and cumulative capacity (upper graph) as a function of their year of initial operation.[22]

(a whopping 4200 MWe) exceeded the BPA's hydroelectric output.[24] Clearly, wind power makes sense there. A counterexample, though, is the small cluster of 18 wind turbines atop Buffalo Mountain in Tennessee. My house is situated such that I get a good view of those wind turbines (although personally, "good view of wind turbines" is another oxymoron). Disappointingly, the turbines rarely operate, especially on those hot muggy days of summer when the roar of air conditioners can be heard throughout our neighborhood. So to repeat myself, the expanded use of renewables is a desirable trend *where it makes sense to do so.*

The most frustrating aspect of wind and solar power is that they are unreliable. As I just mentioned, wind turbines sometimes sit idle when you could most benefit from their output. Similarly, solar power generators are quite ineffective at powering our lights at night, which is my preferred time to use lights. As the portion of wind and solar generators on our grid grows, they bring a commensurate number of challenges to those who manage the grid. The variable and nondispatchable power they produce must be accounted for in other ways, typically by having a comparable amount of on-demand generators such as natural gas plants and a stabilizing amount of base-load capacity. A technology breakthrough for grid-scale energy storage would help to enable the further growth of renewables, but tractable solutions have been elusive so far. If clean base-load capacity is desired, nuclear power is the only serious solution.

The final motivation for the expanded use of nuclear power is nonelectrical applications. In the US, roughly 40% of the total energy consumed is used to produce electricity. Transportation makes up another 30% and 15% is for industrial applications. The contributors to national carbon emissions track a similar distribution since most of the energy comes from the combustion of fossil fuels. Figure 1.5 shows total CO_2 emissions by energy consumption sector and includes the 2050 target (80% reduction relative to 2005).[23] As mentioned earlier, it has been more than 100 years since the US has been at the target level of CO_2 emissions. Even if the US succeeds in completely

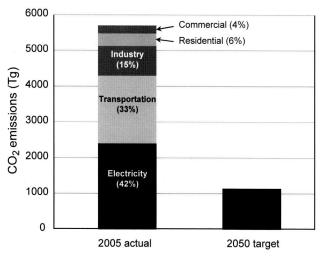

Figure 1.5 Contributions to 2005 US CO_2 emissions and total 2050 goal.[23]

decarbonizing the electricity generation market, it has solved only 42% of the problem. If there is any hope of achieving the national goal, the US must begin to move clean energy generation into the nonelectrical markets. These markets, especially the industrial market, are generally characterized by requiring massive amounts of reliable power on a 24/7/365 basis. The only current energy option that can meet this requirement is nuclear—bar none.

Nuclear power is not completely foreign to nonelectrical applications. Currently there are 59 nuclear plants in 9 different countries that support district heating systems for residential and/or industrial customers. Also, roughly 15 commercial nuclear plants have been used to provide heat to water desalination plants, principally in Japan, India, and Kazakhstan.[25] My Navy colleagues also like to remind me that there are many more nuclear units, albeit not commercial plants, that routinely produce clean water from sea water. However, the relative contribution of nuclear plants for water desalination is quite limited—representing only 0.1% of the global desalination capacity. There are many other industrial applications such as oil refining, plastics and chemicals production, and metals refining that use vast amounts of heat that could be provided by a nuclear plant. There are probably a number of reasons why nuclear plants have lagged integration with industrial plants, including technical, regulatory, and political reasons, but I suspect one of the significant challenges has been the traditional deployment model used for existing large nuclear plants. This hints at one of several potential advantages of smaller-sized nuclear plants, so I will defer a more detailed discussion of nonelectrical applications until a later chapter.

1.4 Challenges for expanding nuclear power

While expanding nuclear power in the electricity sector and introducing it into the nonelectrical energy sectors seems like a no-brainer, there are admittedly a number of challenges. Perhaps the biggest challenge today in the US is the availability of abundant, cheap natural gas. In a complete surprise to most energy analysts, the apparent dwindling role of natural gas in the energy landscape changed dramatically with a new gas recovery technology called hydraulic fracturing, or "fracking" for short. A combination of horizontal drilling and the infusion of high-pressure water into the gas-containing deposits produced a glut of natural gas and caused prices to drop to unprecedented levels. The relative ease (cost and schedule) at which a gas combustion plant can be added to the grid and the more environmentally friendly nature of natural gas, which emits roughly half the carbon per unit energy as coal, makes it a tempting choice for utility executives, despite their having been burned in the past by an overreliance on natural gas only to have the price of their fuel skyrocket.

Although the availability of cheap natural gas presents a challenge to the nuclear industry, I assert that it is a good thing for the US. As mentioned earlier, the US is fortunate to have such a diverse range of natural resources that can all be put to good use. There is already a movement to convert liquefied natural gas (LNG) terminals, originally intended for the import of LNG, into export terminals for the sale of natural gas to countries that are faced with prices that are three to five times higher than in the US. Cheap gas may even entice the relocation of chemical and manufacturing industries back to the US. This is a good opportunity for the country, although I believe it would be more prudent to preserve this natural resource as feedstock for higher value products rather than combust it to produce heat. Although it is a cleaner fuel than coal, it still contributes significantly to national carbon emissions. Furthermore, its relative contribution will grow significantly as coal plants are retired and natural gas becomes the last stronghold of fossil fuel combustion. This leads me to the next challenge for nuclear power or any energy source for that matter: the lack of a sustained national energy strategy.

Earlier I referenced the French experience, in which the country made an explicit decision in 1974 to achieve energy independence through an aggressive implementation of nuclear energy. Today, France generates nearly 80% of its electricity from nuclear power plants and has a healthy electricity export to its neighbors, some of whom have hypocritically rejected nuclear power within their borders. I also briefly mentioned the situation in India, where they are embarking on a three-phase program to develop a sustainable nuclear fuel cycle. What is especially fascinating about the Indian strategy is that every presentation I have heard and every paper I have read from India has echoed the exact same message, regardless of the author or institution. Like the French, they have achieved a single national strategy and are working collectively to achieve it. This is in rather stark contrast to the US, where you are unlikely to hear or read the same vision by any two people. As a result of a lack of national leadership in defining a sustainable energy strategy, companies are reluctant to invest in new technology developments, and utilities are reluctant to invest in anything other than very short-term generating solutions.

Associated with political uncertainty is the investment uncertainty created by an ever-changing political landscape, which has traditionally flip-flopped with changing dominance of political parties. On the plus side, President Obama is the first Democratic president to openly speak in favor of nuclear energy and has done so on many occasions. Unfortunately, the rhetoric does not always reflect political priorities as measured by the distribution of federal investments. Figure 1.6 compares US federal subsidies of several energy sources for both 2010 and 2013.[26] This reality speaks

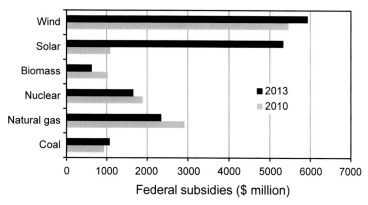

Figure 1.6 US government investments in different energy technologies in 2010 and 2013.[26]

louder than the rhetoric, at least to investment and user communities. Also, the pronuclear rhetoric does not explain the indefensible abandonment of billions of tax-payers' dollars that have been spent over the past decades to provide a permanent repository for high-level nuclear waste. The result is a confusion of mixed signals to the nuclear industry, the investor community, and the general public.

Related to the fluctuating political support of nuclear power are the numerous policies favoring other specific energy forms, especially renewables, in ways that greatly distort energy markets. These include such things as tax credits for adding renewable generators and mandated energy portfolios that require grid dispatchers to preferentially take electricity from wind turbines or solar farms in spite of potential grid instabilities that they create. It was federal tax advantages that encouraged our local utility in Tennessee to build the wind turbines on Buffalo Mountain, despite a capacity factor of less than 20%. The result of these market distortions is that some regions of the country actually experience negative electricity pricing, that is, more electricity is produced than can be sold. This places established base-load plants at a disadvantage, and we have already begun to see the permanent closures of high-performing baseload plants due to poor market economics, including the Kewaunee nuclear plant in 2013 and the Vermont Yankee plant in 2014.

Switching away from politics, another major challenge for nuclear power, exacerbated by the national and global financial crisis triggered in 2008, is the total price tag for a modern nuclear power plant. While it is difficult and complex to know the exact cost of a nuclear plant, a figure that is held privately between the seller and buyer, speculated values range from $5–8 billion US dollars for a gigawatt-class plant. The two-unit plant being constructed at the Vogtle site in Georgia has been reported to cost a teeth-jarring $14 billion. In the US, this means that only the largest utilities can even consider such a purchase. This forbidding cost challenge is largely a consequence of the fact that nuclear plants on the market are all gigawatt-class plants, some approaching nearly 2 GWe. During a testimony to the Congressional Committee on Environment and Public Works in 2007, Vice President Al Gore stated the following: "The problem is these things (nuclear power plants) are expensive. They take a long time to build, and at present they only come in one size, extra large."[27] After the very first commercial plants constructed in the late 1950s, plant supersizing has been the national and global trend. Supersizing, coupled with custom-made and complex plants, added to the costs through increases in materials and labor and the carrying costs of the financing during lengthy construction schedules. The result is that the nuclear solution is not a viable option for many utilities and energy customers who cannot afford a big plant and who do not need that amount of generating capacity. This is equivalent to going to the local car lot to replace the family sedan and finding only Hummer H1s. As much as I'd like to take one home, it is not a viable option for me and many others.

Finally, another significant challenge comes as a result of the three high-profile accidents at nuclear plants: the US Three Mile Island plant in 1979, the Russian Chernobyl plant in 1986, and the Japanese Fukushima Daiichi plant in 2011. The first immediate impact of these accidents was an erosion of public opinion, causing some entire countries to abandon nuclear power. Fortunately, public support rebounded after each accident and is now solidly favorable in the US and several other countries. A more enduring challenge resulting from these accidents is due to the inevitable response of the nuclear industry and regulator. On the one hand, lessons-learned have provided valuable insight and guidance for improving the technology and engineering of nuclear plants and have added substantially to the safety and security of all existing and new plants. On the other hand, the accidents have contributed to a progressive escalation of regulatory requirements for designing, constructing and operating nuclear power plants with a commensurate increase in cost and complexity, especially for existing plants that must retrofit the required changes.

These collective challenges have all contributed to a disappointing slowdown of the nuclear renaissance that looked so promising in the early 2000s. Of the 18 new combined license applications filed with the US Nuclear Regulatory Commission, eight have been suspended or withdrawn as of September 2014. Although five new plants are under construction in the US, five existing reactors were permanently closed or announced to be closed in 2014. While three of these were for technical/political reasons, two plants were closed due to the badly distorted energy market in their locations that penalized base-load plants. It seems as though the nuclear renaissance is destined to the timescale of the European Cultural Renaissance, which evolved over nearly four centuries.

Enter the SMR—desperate hope or game-changer?

1.5 New interest in small nuclear power

As will be discussed in the next chapter, smaller-sized commercial nuclear reactors have been around for several decades. They are generally characterized as having power outputs of less than 300 MWe, in contrast to large plants that have power outputs exceeding 1000 MWe. They are also typically designed so that the entire nuclear steam supply system (reactor unit) can be prefabricated in a factory and transported to the site, where it is installed and operated with multiple identical units. Despite their lengthy history, it has been only recently that SMRs have dominated the industry rhetoric. Why? And why now?

I will speak to the "why?" more thoroughly in Part Two, Fundamentals and features. In short, they offer a host of benefits, especially related to safety, affordability, and flexibility of use. The answer to "why now?" lies in the convergence of several factors. First and foremost, we would not be talking about SMRs if it were not for the excellent performance and safety record of the existing fleet of commercial nuclear power plants. Unlike the 1970s, when the collective capacity factor for US nuclear plants bounced around the 50% level, the fleet average has been at 90%, plus or minus, for the past 15 years. And although some may debate the public perception about the safety of nuclear facilities, in reality, they have surpassed all other industries in virtually every safety metric. Without this phenomenal performance record, politicians would not be speaking in favor of nuclear power, and utility executives would not be considering the newest generation of nuclear plant designs.

Second, the more serious rhetoric about cleaning up the energy mix has placed more attention on nuclear in general, but especially SMRs because of their apparent suitability as a potential one-for-one replacement of aging coal plants that are being programmed for closure over the next 10–20 years. An overlaying factor is the national and global economic crisis that hit in 2008, which had the effect of reducing investor risk tolerance. The enormous price tag for gigawatt-class nuclear power plants has created a financing nightmare that may be eased considerably with the smaller and staggered capital cost liabilities with SMRs.

A domestic interest that favors SMRs in the US is the fact that the country has mostly given away the large plant business. The traditional vendor giants include Westinghouse, which is now owned by Toshiba, and General Electric, which is partnered with Hitachi. For all large plants, major components such as the reactor and steam generator vessels can only be manufactured overseas where the component suppliers still exist for local nuclear markets. A national goal for domestically designed SMRs is that they also be manufactured in the US, initially using existing manufacturing capacity and eventually with dedicated factories. This will help to bolster the domestic economy, preserve nuclear resources and expertise, and allow the US to be highly competitive in the international market.

The factors discussed above contribute to the final factor that has brought attention to SMRs in recent years: their suitability for nontraditional energy markets by virtue of their smaller unit size, expanded safety margins, and flexible plant designs. These nontraditional markets include both the small-demand electricity markets, currently served by coal plants or even diesel-fired burners, and nonelectrical markets where the valued product is heat. SMRs are also well suited for cogeneration markets where both electrical and process heat production can be integrated with other industrial applications.

All of these factors will be explored in more detail in later chapters. But first, I will back up and provide a brief history of SMRs, which is intimately woven within the broader history of nuclear power from its very beginning. Chapter 2 covers the first half century of nuclear power from roughly 1950 through 2000, while Chapter 3 covers SMR activities from the year 2000 through today.

References

1. Pasternak AD. *Global energy futures and human development: a framework for analysis*. Lawrence Livermore National Laboratory; October 2000. UCRL-ID-140773.
2. *Human development report 2014*, United Nations Development Programme. Available at: http://hdr.undp.org/en/content/human-development-index-hdi.
3. Electric power consumption (kWh per capita), The World Bank. Available at: http://data.worldbank.org/indicator/EG.USE.ELEC.KH.PC.
4. Ouedraogo NS. Energy consumption and human development: evidence from a panel cointegration and error correction model. *Energy* December 2013;**63**:28–41.
5. Anderson MT, Woosley Jr LH. Water availability for Western United States—key scientific challenges. US Geologic Survey. *Circular* 1261, 2005.

6. *The great Georgia-Tennessee border war of 2013 is upon us.* The Atlantic Wire; March 25, 2013. www.theatlanticwire.com/national/2013/georgia-tennessee-boarder/63508.

7. Global Water Intelligence: Desalination.com, www.desalination.com/market/desal-markets, September 18, 2013.

8. *Energy demands on water resources: report to congress on the interdependence of energy and water.* US Department of Energy; December 2006.

9. *International Energy Outlook 2013*, US Energy Information Administration; 2013.

10. McFarlane H, et al. *American Nuclear Society Mission to India.* Idaho National Laboratory; March 2007. INL/MIS-07–12356.

11. *Nuclear power in China*, World Nuclear Association. Available at: http://www.world-nuclear.org/info/Country-Profiles/Countries-A-F/China.

12. Emerging nuclear countries, World Nuclear Association. Available at: http://www.world-nuclear.org/info/Country-Profiles/Others/Emerging-Nuclear-Energy-Countries/.

13. Remarks by the President, Office of the White House. Available at: http://www.whitehouse.gov/the-press-office/2011/01/24/remarks-president-state-union-address.

14. *Comparison of lifecycle greenhouse gas emissions of various electricity generation sources*, WNA Report. Available at: http://www.world-nuclear.org/WNA/Publications/WNA-Reports/Lifecycle-GHG-Emissions-of-Electricity-Generation/.

15. *Frequently asked questions: what is US electricity generation by energy source?* US Energy Information Administration; April 2015. www.eia.gov/tools/faqs.

16. Tucker W. *Terrestrial energy.* Washington, DC: Bartleby Press; 2008.

17. Cravens G. *Power to save the world: the truth about nuclear energy.* New York: Alfred A. Knopf; 2008.

18. Weinberg AM, Spiewak I, Barkenbus JN, Livingston RS, Phung DI. *The second nuclear era.* Praeger Publishers; 1985.

19. *Federal leadership in environmental, energy, and economic performance.* Office of the White House; October 2009. Executive Order 31514.

20. Net generation by energy source, Energy Information Administration. Available at: www.eia.gov/electricity/annual/html/epa_01_02.html.

21. *National CO_2 emission from fossil fuel burning.* Carbon Dioxide Information Analysis Center, Oak Ridge National Laboratory; May 2009.

22. Annual electric generator report, form EIA-860, Energy Information Administration. Available at: www.eia.gov/electricity/data/eia860.

23. *Inventory of US greenhouse gas emissions and sinks: 1990–2012.* US Environmental Protection Agency; April 2014. EPA 430-R-14–003.

24. *Wind power surpasses hydro for the first time in the Northwest region*, OregonLive, Available at: http://www.oregonlive.com/environment/index.ssf/2012/10/wind_power_surpasses_hydro_for.html.

25. *Advanced application of water-cooled nuclear power plants.* International Atomic Energy Agency; July 2007. TECDOC-1584.

26. *Direct federal financial interventions and subsidies in energy in fiscal year 2013.* US Energy Information Administration; March 2015. www.eia.gov.

27. Gore A. *"Vice president Al Gore's perspective on global warming," hearing before the committee on environment and public works, 110th congress.* March 21, 2007.

A brief history of small nuclear power (1950–2000)

2

I never liked or appreciated history classes in high school. I remained disinterested in history until a few years ago when I overheard some young coworkers discussing a historical event, an event that I vividly recall experiencing firsthand. It was then that I realized that I was, and continue to be, a part of history. History suddenly got interesting.

A lot has been written about the history of nuclear energy. Whole books on the subject have emerged as a number of industry pioneers are now retiring and spending their free time reminiscing about the many remarkable early accomplishments of the industry. I will not attempt to replicate those fascinating stories here, but clues to the possible future of small modular reactors (SMRs) may be found in the industry's past. And although the focus of this history lesson is on smaller-sized power systems, with a primary focus on the US experience, I cannot help but include nuggets from the history of large-sized reactors. Their histories, large and small, are inextricably connected. In many ways, SMRs are motivated by both the successes and the failures of traditional large nuclear plants.

2.1 Military propulsion and power

Commercial nuclear power in the US grew out of the successful development and deployment by the US Navy of small reactor systems for marine propulsion—initially for submarines and later for surface vessels. The first nuclear-powered submarine was the USS Nautilus launched in 1954. Six years later, the first nuclear-powered aircraft carrier was launched: the USS Enterprise. The Nautilus operated for 26 years and was decommissioned in 1980, while the Enterprise remained in service for a remarkable 52 years until being retired from service in 2012.

When Captain Hyman Rickover, later to become Admiral and the undisputed father of the US Nuclear Navy, embarked on the construction of the Nautilus in 1949, he contracted two parallel construction projects: one with Westinghouse to construct the Nautilus, powered by a small water-cooled reactor, and one with General Electric to construct the Seawolf, powered by a sodium-cooled reactor. The Nautilus was completed and put to sea 3 years ahead of the Seawolf. The Seawolf experienced a number of maintenance issues, the most serious of which was due to sodium-steel incompatibilities in the steam superheater units. After 2 years of operations, the Seawolf's sodium-cooled reactor was replaced with a water-cooled reactor—the one that had been built as a spare for the Nautilus.[1] Historians, especially those who had worked for Westinghouse or General Electric, differ in their recollections of why Rickover selected light-water reactor technology as the basis for his future fleet of naval vessels.

Small Modular Reactors. http://dx.doi.org/10.1016/B978-0-08-100252-0.00002-1

Perhaps it was the substantial influence of Alvin Weinberg, who is generally considered to be the father of light-water reactors and had helped to train Rickover and his staff on nuclear technology. Maybe it was the sodium-steel incompatibilities that plagued the Seawolf. Or maybe it was that fact that sodium metal tends to ignite rather spectacularly when contacting water, which would not seem to be a logical choice for a machine intended to be submerged in sea water for several months at a time. Whatever the reasons, water-cooled reactors became the propulsion unit of choice for the Navy and consequently set the direction of the commercial nuclear power industry for decades to follow—small and large.

The Navy Nuclear Power Program continues to be highly successful with a stellar record of performance and safety. Despite their proven success for the Navy, the design of these SMRs would require significant modifications to be adapted for commercial power production. In addition to being tightly controlled information, their design details, fuel form, and materials of construction were all selected to meet very specific and demanding performance goals that are quite different than the goals of commercial power plants. However, the vast design, manufacturing, and operational experience of the nuclear Navy is not entirely lost on the budding commercial SMR industry. First, many of the engineers and managers that fill the ranks of vendors, suppliers, and commercial nuclear plants originally cut their teeth in the nuclear Navy. Second, the mature and highly refined expertise of the Navy's manufacturers and component suppliers are being tapped directly by many of the current SMR vendors. Electric Boat, Newport News, Babcock & Wilcox, and Rolls Royce (supports the UK Royal Navy) are some of those dual-purpose manufacturers engaged in commercial SMR development. Finally, some of the current SMR designs being developed in other countries, most notably the Russian Federation, have evolved directly from the design of their marine propulsion systems.

Propelled by the successful utilization of nuclear energy by the Navy, the US Air Force and the US Army also embarked on nuclear power programs. Although much less enduring than the Navy's program, the Air Force and Army programs resulted in considerable learning and technology development pertinent to commercial nuclear power, especially SMRs. Many of the motivations for their consideration of nuclear power have striking parallels with contemporary motivations for pursing commercial SMRs today.

Beginning in 1946, the Air Force explored the use of small nuclear reactors to power long-range bombers as part of the Aircraft Nuclear Propulsion (ANP) program.[2] The US was in the throes of the cold war with the Soviet Union, and it was highly enticing to have a bomber with the capability to stay airborne for long periods of time, able to reach enemy territory without refueling. A total of six small reactors were constructed during the course of the program, which involved several national laboratories and industry partners. Specifically, two research reactors and one test reactor were built at Oak Ridge National Laboratory (ORNL) in Tennessee and three high-temperature prototype reactors were built at the National Reactor Testing Station in Idaho. I have a sentimental attachment to one of those research reactors: the ORNL Tower Shielding Reactor. Early in my career at ORNL, I assumed responsibility for the operations and experiments at the Tower Shielding Facility (TSF) and was privileged to work with talented engineers and experimentalists such as Leo Holland and Francis ("Buzz") Muckenthaler. Although the Tower Shielding Reactor was constructed to develop and confirm the crew compartment shielding for the ANP program, the unique spherical design of the reactor proved to be valuable for a large array of shielding investigations that supported virtually every advanced reactor program conducted during the subsequent three decades.[3]

Near the conclusion of the ANP program, a Convair B-36H bomber was converted to include an operating 3 MW reactor, although the reactor was not used to propel the aircraft. Using the TSF to test and validate candidate shield designs, a lead- and rubber-shielded crew compartment was developed to protect the flight crew from the reactor's radiation. The prototype aircraft accumulated over 200 flight hours in nearly 50 flights over Texas and New Mexico in the late 1950s prior to the cancellation of the program.

A few years ago, I had the opportunity to lecture on the history of SMRs at a Rotary Club group in Oak Ridge, Tennessee. As you might guess, Oak Ridge is home to ORNL, a lab that played prominently in the ANP program. I was approached after my lecture by a white-haired gentleman who had worked on the ANP program and offered me an elaboration on a photo that I had included in my talk: an aerial shot of the NB-36H bomber and its accompanying escort plane,[4] which is reproduced in Figure 2.1. This soft-spoken engineer explained to me that the escort plane was always present during flight tests and contained several marine paratroopers. In the event of a malfunction of the reactor aboard the bomber, the standard procedure was to drop the reactor through the bomb doors. Simultaneously, the marines would parachute to the crash site and secure what was left of the reactor until ground support arrived. In hindsight, this procedure may not have been such a great idea. Fortunately, it was never required.

Figure 2.1 Aerial photograph of NB-36H aircraft with an operating reactor onboard.[4]

Surprisingly it was not the questionable merits of the application that terminated the ANP program in 1961 after an investment of over $1 billion (in 1960s dollars).[5] Rather, its cancellation has been attributed to a variety of reasons, including significant progress in conventional aircraft propulsion, the advent of the intercontinental ballistic missile, and President Kennedy's interest in redirecting funds to race the Soviets to the moon. An article in *Scientific American*[6] suggested that it might be time to resurrect the nuclear airplane to help clean up the skies. True, the thousands of commercial airline flights that crisscross the US every day contribute measurably to total carbon emissions. And as a frequent passenger on those airliners, I would love to hop coast-to-coast a few minutes faster. But I will offer the same advice that I levied at wind and solar energy sources in the previous chapter. We should only use nuclear power *where it makes sense to do so*. Nuclear-powered air travel does not.

The Army Nuclear Power Program ran between 1954 and 1976 and resulted in the construction of eight reactors. According to Lawrence Suid, who was contracted by the Army to compile a history of their program,[7] the reactors included five 1–2 MWe pressurized water reactors (PWR), which were adapted directly from Rickover's submarine units, one 1 MWe boiling water reactor (BWR), one 10 MWe barge-mounted PWR, and one 0.5 MWe gas-cooled reactor (GCR). These eight reactors are listed in Table 2.1. The Army's program was eventually discontinued due to poor economics of the nuclear plants relative to cheaper alternative fuels and shifting national priorities. But like the nuclear power programs of the Navy and Air Force, it pioneered many innovations in nuclear power.

The Army's nuclear power program was motivated in part by the nuclear power frenzy that was sending large budgets to the Navy and Air Force. Although there are many stories of cooperation among the services as they collectively expanded the use of nuclear power, there are also clear examples of interservice competition. One of the more amusing anecdotes shared by Suid in his recount of the Army's program was regarding the mobile unit, the Sturgis, operated on a floating barge. The Sturgis was not built from scratch but rather was a major overhaul of the Charles H. Cugle liberty

Table 2.1 **Reactors built by the US Army Nuclear Power Program**[7]

Reactor	Power (MWe)	Type	Location	Startup	Shutdown
SM-1	2	PWR	Fort Belvoir, Virginia	1957	1973
SM-1A	2	PWR	Fort Greely, Alaska	1962	1972
PM-1	1	PWR	Sundance, Wyoming	1962	1968
PM-2A	1	PWR	Camp Century, Greenland	1960	1962
PM-3A	1.5	PWR	McMurdo station, Antarctica	1962	1972
SL-1	1	BWR	Arco, Idaho	1958	1960
MH-1	10	PWR	Panama Canal (Sturgis)	1967	1976
ML-1	0.5	GCR	Arco, Idaho	1961	1966

ship, which sported an operable diesel propulsion unit. The Army decided to remove the diesel unit and convert the ship to a barge to avoid the possibility that Adm. Rickover would claim it for his nuclear Navy.

The Army rightfully saw nuclear power as an opportunity to provide power to remote installations where resupply of traditional fuels was difficult and expensive. It could provide those installations with highly reliable power for multiple years on a single charge of fuel. Remember those statements—you will have an overwhelming sense of déjà vu in later chapters when I review some of the contemporary motivations for SMRs. As frequently happens, though, the Army did not stop with this sensible application for nuclear power and instead conducted multiple studies on the use of small reactors to power trains, large overland haulers, and yes, even nuclear-powered tanks. I am reminded of some fatherly advice that I offered frequently to my teenage daughters: "Just because you *can* do something doesn't mean that you *should*." The thought of a nuclear-powered tank, which is basically a mobile battlefield target, gives me chills. Fortunately, the Army's Transportation Corps had a similar reaction and did not pursue it further.

The Army succeeded in demonstrating the construction and operation of small reactors in very harsh environments. By the time they were ready to deploy a reactor to Antarctica, they had learned an important lesson: design the plant to be modular, and minimize the amount of on-site construction in order to reduce costs and shorten the deployment schedule. This lesson has not been lost with the demise of the Army's nuclear power program—it is a central feature of many SMR designs being developed today. The program also accomplished a number of nuclear "firsts," such as the first use of stainless steel as fuel cladding, the first in-place annealing of a reactor vessel, the first land-transportable nuclear reactor, and the first use of nuclear power to desalinate water.

2.2   Commercialization of nuclear energy

Commercial marine propulsion was a logical successor to the Navy's application of nuclear power. In his famous "atoms for peace" speech in 1955, President Eisenhower proposed the construction of the nuclear-powered NS Savannah. The Savannah, which was launched in 1962 and powered by a 69 MWt PWR, was first and foremost a demonstration of a peaceful use of nuclear power. Its stylish appearance made it look more like a cruise ship than a merchant hauler. Before being retired from service in 1971, the showboat visited over 70 domestic and foreign ports. Three other nuclear-powered commercial merchant ships were constructed: the German-built Otto Hahn, the Russian-built Sevmorput, and the Japanese-build Mutsu.

The 38 MWt reactor that powered the Otto Hahn is of special interest because it was the first commercially deployed example of an integral PWR. The integral configuration, in which all primary system components are contained within a single vessel, is used by many SMR designs, old and new, due to the system simplicity and enhanced safety that it allows. The features and benefits of this design arrangement are discussed thoroughly in Part Two.

The small fleet of nuclear icebreakers built by Russia is in a special category of marine propulsion that is neither for military missions nor for commercial applications. These small reactor plants with nominal power of 100–200 MWt greatly extend the navigable season of the Arctic Ocean. They also serve as a basis for some of Russia's entries in the SMR market.

The early land-based reactors for commercial power production were commissioned in the late 1950s and early 1960s and were essentially scaled-up versions of naval PWR plants. The 60 MWe Shippingport plant began operation in 1957, the 200 MWe Dreseden plant in 1960, and the 250 MWe Indian Point Unit 1 plant in 1962. The 5 MWe Vallicetos plant, which began operation in 1957, was owned by General Electric and served as a demonstration plant for BWRs. Although all of these plants were "small" by today's terminology, they were designed to be scalable to very large sizes and built to give the fledgling commercial nuclear power industry experience with the construction and operation of nuclear plants.

Buoyed by the rapidly growing demand for electricity in the US, the high level of confidence in the safety of nuclear plants (based largely on the Navy's experience), and the economic principle of "economy of scale," utilities raced to order larger and larger plants. The largest of the US plants eventually exceeded 1300 MWe—more than 10 times the size of the first demonstration plants. Figure 2.2 shows the progression of power capacities for the commercial nuclear power plants built in the US.[8] Note that most of the plants commissioned before 1970 had capacities below 300 MWe while

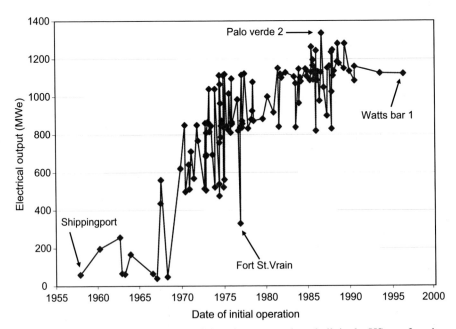

Figure 2.2 Electrical output of commercial nuclear power plants built in the US as a function of their initial date of operation.[8]

all plants built after 1970 had capacities greater than 500 MWe. The largest capacity plant built in the US was the 1335 MWe Palo Verde Unit 2, and the last plant to go into operation was the 1121 MWe Watts Bar Unit 1. The one anomaly in the growth trend was the startup of the demonstration GCR, Fort St. Vrain, in 1976. No subsequent GCRs were built in the US, although they continue to be pursued. Internationally two small modular GCRs are being constructed in China.

Spurred by the momentum of a postwar manufacturing engine and with the promise that nuclear power would be "too cheap to meter," the fledgling nuclear industry grew at an incredible rate during the 1960s. By 1970, there were four major nuclear plant vendors in the US: Westinghouse, General Electric, Babcock & Wilcox, and Combustion Engineering. Competition for plant orders was fierce and prompted vendors to offer fixed price, "turnkey" bids. By the end of 1967, US utilities had ordered 75 new nuclear plants, 60 of which were ordered in 1966 or 1967. The combined generating capacity of these plants was more than 45,000 MWe.[9]

The oil embargo in 1973 by the Organization of Petroleum Exporting Countries further fueled the nuclear power frenzy. Crude oil prices more than doubled, and within 2 years, domestic coal prices also doubled due to the high demand on coal as a replacement for scarce oil. This fossil fuel energy crisis seemed to convince even the nuclear skeptics that nuclear power would eventually dominate the energy market.

The 1973 energy crisis, combined with the apparent but yet unproven success of the large nuclear plants, gave rise to several studies between 1974 and 1977 on the use of nuclear power for heat applications rather than for electricity generation. Industrial applications at that time constituted about 40% of the total US energy consumption, supplied entirely by fossil fuels (51% natural gas, 27% oil, and 22% coal)—similar to today's energy demographics. The initial study, conducted by ORNL, concluded that coal should be used to replace high-priced oil and natural gas, and that nuclear power was a viable energy source for industrial applications.[10] The study also concluded that smaller-sized nuclear units were a better match for the energy demands of industrial plants. Several studies of process heat applications by NuScale Power have resulted in the same conclusion today.[11]

The ORNL study spurred several follow-up studies funded by the US Energy Research and Development Agency (ERDA), which was the predecessor to the Department of Energy, to explore smaller nuclear reactor concepts. These small reactor designs were summarized in a later ORNL report[12] and included the following:

- a small (300–400 MWt) industrial energy PWR derived from the Babcock & Wilcox Consolidated Nuclear Steam Generator (CNSG) reactor, which had been the basis for the nuclear unit that powered the NS Otto Hahn merchant ship;
- a 300–400 MWe small BWR concept developed by General Electric based on their earlier prototypes;
- a 200–1000 MWt gas-cooled pebble-bed reactor concept also developed by General Electric and based on the German 40 MWt AVR test reactor; and
- an 850 MWt GCR derived from the larger General Atomics Fort St. Vrain design.

Additional ORNL and Babcock & Wilcox studies evaluated a 91 MWe (365 MWt) land-based version of the CNSG for process energy (PE-CNSG).[13] At the request of the

ERDA, Babcock & Wilcox also developed a larger version, the 400 MWe (1200 MWt) Consolidated Nuclear Steam Supply (CNSS) reactor. Both the CNSG and the CNSS designs utilized integral PWR configurations similar to the propulsion unit in the Otto Hahn. The first PE-CNSG plant was targeted for a sulfur mining operation in Texas, but the project was short-lived as dark clouds began to form quickly over the nuclear industry in the mid-1970s. In 1979, Babcock & Wilcox management decided not to pursue further development of the CNSG and CNSS designs due to a growing antinuclear community.

2.3   Exuberance to exasperation

Although nuclear power appeared to be destined to dominate the energy market in 1973, conditions were already at play that would completely reverse that perspective within the following few years. Many believe that it was the accident at the Three Mile Island (TMI) plant in 1979 that halted the expansion of nuclear power in the US, but in reality, many factors contributed to the dramatic reversal in the expanding nuclear industry. One year prior to the TMI accident, Bupp and Derian published a book entitled *Light Water: How the Nuclear Dream Dissolved*,[14] in which they suggest that the roots of the demise of the US commercial nuclear industry began in its very infancy.

A significant contributing factor to the industry's demise was the rapid scale-up of plant size in a very short span of years. Much of this scale-up occurred over a relatively short 10-year period between 1960 and 1970 after only minimal operating experience with the much smaller prototypes. In Figure 2.2, I showed the capacity of new plants brought online during this time period. A very revealing presentation of related data is shown in Figure 2.3, which compares the size of plants being ordered and the size of plants actually in operation. The 10-year lag time in the design, engineering, licensing, and construction of new plants created a situation in which utilities were ordering (and vendors were selling) plants that were sixfold larger than current operational experience. This compares to a more traditional rule of thumb that recommends a twofold scale-up for complex engineered systems. The more aggressive approach employed by the nuclear industry required a significant leap of faith, which unfortunately did not play out well.

As plant sizes grew and operational issues began to moderate the industry's confidence in the ultimate safety of the plants, more stringent safety and environmental requirements were imposed, and the elegant simplicity of the original light-water reactors gave way to a complex layering of redundant safety and auxiliary systems. This escalation of plant complexity contributed to rapidly increasing costs, licensing delays, construction and operational delays, and eventually decreased confidence by the owners and lenders in the profitability of the plants. The prevailing approach to design each new plant as a "one of a kind" to accommodate the customized interests of individual customers also contributed to increased licensing, construction, and operational complexities. The result was that after 20 years of plant design and construction experience, the price of nuclear plants continued to increase and remain highly uncertain.

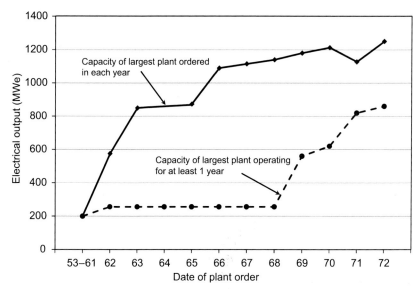

Figure 2.3 Capacity of the largest plant ordered in a given year compared to the capacity of the largest plant operating for at least 1 year.[8,14]

These many challenges gave abundant fodder to nuclear opponents, who were able to cast doubt regarding the viability of commercial nuclear power. By the end of the 1970s, it became unwise for politicians to be openly supportive of nuclear power. Then on March 28, 1979, a major accident occurred at the TMI plant in Pennsylvania. It was the worst accident to occur at a US nuclear plant and resulted in a partial meltdown of the reactor core.[15] Although relatively benign by normal standards for industrial accidents, the TMI accident served to punctuate the demise of the once exuberant nuclear industry. Adding insult to injury was the fact that the 1973 oil embargo, while seeming to have bolstered support for nuclear power, caused the general population to reduce its energy consumption in the ensuing years in response to higher electrical bills. The result was a substantial overbuild of generating capacity and a doubling of the capacity margin from 15% in 1970 to nearly 30% by 1980. So even if the nuclear plants did perform well, they simply were not needed. Finally, very high interest rates during the mid- to late 1970s created a financial barrier and caused the cancellation of many proposed power plants, both nuclear and coal.

In addition to the rise and fall of light-water reactors, a parallel development was occurring for another type of reactor: the fast-spectrum reactor. The rapid growth of nuclear power in the 1960s and early 1970s led to concern that we would soon exhaust our supply of uranium. This prompted the development of fast-spectrum reactor designs that could breed more fuel than they consumed. A substantial program ensued to develop and demonstrate the liquid-metal-cooled fast-breeder reactor. The 400 MWt Fast Flux Test Facility began operation in 1980, and fabrication of the 375 MWe Clinch River Breeder Reactor demonstration plant was 80% completed before the project was terminated in 1984, due largely to political concerns of nuclear weapon proliferation. Prior to the cancellation of the Clinch River Breeder Reactor, the widely held vision was that large (>1000 MWe) metal-cooled fast reactors would eventually replace the existing fleet of large water-cooled reactors.

2.4 Redirecting the nuclear industry

In 1982, with the nuclear industry "on the ropes," the Electric Power Research Institute (EPRI) conducted a survey of management, operations, and maintenance personnel from 11 utilities that operated commercial nuclear plants and gleaned a number of interesting findings regarding the perceived safety and operability of the existing plants.[16] Of special relevance to SMRs were the following findings from their study: (1) incrementally improved post-TMI light-water reactors pose very low risks to the public but high risks to investors, (2) 1200–1300 MWe plants were felt to be too large and complex, (3) the plants respond too rapidly to transients, and (4) nuclear plants need to be less sensitive to events in the secondary systems. Most contemporary SMR designs seek to avoid these undesirable traits through a variety of deliberate design choices, as will be discussed in Chapter 5.

Spurred in part by the conclusions from the EPRI survey, Alvin Weinberg and colleagues at the Institute for Energy Analysis conducted a study on the feasibility of designing a reactor that is intrinsically safe under all conceived operational circumstances.[17] Weinberg was a pioneer in the early development of nuclear power, beginning with the Manhattan Project. He served as the director of ORNL between 1955 and 1973, a period of extensive exploration of nuclear power that included reactor designs ranging from traditional light-water reactors to exotic designs using molten salts and circulating fluid fuel. Although Weinberg held the original patent on PWRs, he became interested in more robust designs and technologies. While no design available at the time of the Institute's study met all of the safety criteria, the 400 MWe Swedish-designed Process Inherent Ultimately Safe (PIUS) concept appeared to achieve the goals the best, followed by the US-designed 100 MWe modular high-temperature gas-cooled reactor (MHTGR). Weinberg concluded that it was possible to design a nuclear plant to be sufficiently forgiving, but he had considerable concern for its economic viability. Weinberg stated the following in his characteristically crisp manner: "Unless a forgiving reactor is affordable, no one will buy it." We continue to face that reality today—the subject of Chapter 6.

The PIUS reactor was a unique design that achieved extreme safety margins through an innovative arrangement of the reactor internals such that the hot primary reactor coolant circulated within a larger pool of cold borated water.[18] A static hot/cold water interface maintained separation of the two pools during normal operation. In an accident situation, a hydraulic imbalance would occur, and the cold borated water would enter the primary coolant system to shut down the reactor and remove decay heat using only natural forces and without requiring operator action. The other significant safety feature of PIUS was the very thick prestressed concrete reactor pressure vessel that contained the primary system and the borated pool. Although larger than modern SMRs, which mostly have capacities below 300 MWe, the PIUS design influenced subsequent SMR designs through its innovative approach to ensuring long-term cooldown of the reactor.

The MHTGR was downsized from a 1200 MWe design specifically to make the design "inherently" safe against severe accidents.[18] This performance is achieved through a combination of a very robust fuel form and a geometric arrangement of the reactor core that ensures adequate conductive heat removal. The fuel is made from tiny uranium kernels that are coated in multiple layers of carbon and silicon carbide, allowing it to maintain fuel integrity up to very high temperatures. By configuring the fuel/moderator blocks in an annular region, residual decay heat in the core can be effectively conducted to the reactor vessel and ultimately to the ground even if primary coolant is lost.

The EPRI study also spurred the Advanced Light-Water Reactor program. In the early part of the program, utilities and vendors cooperated in developing a comprehensive set of standard design requirements for new plants in the hopes of avoiding the one-of-a-kind nature of the existing plants. These user requirements were directed at developing a standardized large (nominally 1200 MWe) plant with increased design margins and robustness. This led to two evolutionary designs: the General Electric Advanced Boiling Water Reactor (ABWR) and the Combustion Engineering System 80+. Both designs eventually received design certification from the US Nuclear Regulatory Commission (NRC), although neither has been ordered or built in the US. However, the ABWR was eventually adopted by Japan, and the System 80+ was adopted by the Republic of Korea.

The Advanced Light-Water Reactor program also instigated the development of two smaller plant designs that incorporated "passive" safety features, that is, features that rely on fundamental laws of physics (gravity, natural circulation, and stored energy) for operation rather than engineered systems. These included a 600 MWe PWR and a 600 MWe BWR. Westinghouse led the development of the small PWR, designated as the Advanced Passive (AP-600) design, while General Electric led the development of the small boiling water, designated as the Simplified Boiling Water Reactor (SBWR). In addition to using passive safety features, both designs incorporated design simplifications that could reduce the cost of the plant by reducing the amount of bulk material (concrete and steel) and components (valves, pumps, and pipes).

The AP-600 was subsequently up-sized to the AP-1000 design with a power capacity of 1140 MWe. The SBWR was also subsequently super-sized to the 1500 MWe ESBWR, making it one of the largest reactors on the market. Both the AP-1000 and the ESBWR eventually received NRC certification, although only the AP-1000 has been ordered to date. Interestingly, the AP-1000 is the smallest of the nuclear plants being marketed in the US and is also the design that has the greatest number of identified potential customers. As of September 2014, half of the active applications for combined operating licenses filed with the NRC were for AP-1000 units with the other half being distributed among three other reactor designs.[19] I would like to say that it is the AP-1000's relatively small power capacity compared to the ESBWR that has made it a popular choice, but it may also be the fact that it received NRC certification in 2005 while the ESBWR design was not certified until 2014.

In parallel to the Advanced Light-Water Reactor program, the government initiated the Advanced Liquid-Metal Reactor program, involving a partnership of industry and laboratories to develop new fast-breeder reactor designs that made greater use of passive or inherent safety systems. The result was a design run-off between two smaller-sized sodium-cooled reactors: General Electric's Power Reactor Inherently Safe Module (PRISM) and Rockwell's Sodium Advanced Fast Reactor. The PRISM design was ultimately selected for further development and was unique in its operational philosophy of using nine small (160 MWe) power modules to comprise a 1400 MWe plant—comparable to the largest Light-Water Reactor units on the market. As will be described later, the extensive use of passive safety features and the multimodule plant model that was used for PRISM has been carried forward to several of the SMR designs under development.

A personal note: I trace my enthusiasm for SMRs back to PRISM. During the Advanced Liquid-Metal Reactor program, ORNL partnered with General Electric to provide analysis support to design the in-vessel shielding for PRISM. I was assigned to the project and was immediately attracted to the notion of modularizing the nuclear system. Having analyzed only large plant designs previously, the simplicity and flexibility of the PRISM SMR was a refreshing and compelling departure from traditional thinking.

2.5   Early international SMR activities

As the US nuclear power program began winding down in the 1980s and began to re-create itself through the advanced reactor programs, the rest of the world continued to embrace nuclear power, while also casting an anxious eye toward what had happened in the US. Most countries with mature economies already had a good start on commercial nuclear power and countries with emerging economies were actively pursuing it. Heisling-Goodman published a paper in 1981 detailing the conditions and power requirements in those emerging countries and concluded that small power reactors were the sensible solution.[20] Installed grid capacity was a major consideration since 41 of the 65 developing countries evaluated in their study had total grid capacities less than 1 GWe. The dilemma for these countries was that the established reactor vendor base was focused on selling large plants to large electrical markets. With the shelving of Babcock & Wilcox's CNSG and CNSS designs, the primary vendors promoting smaller designs suitable for developing countries included Rolls Royce of the United Kingdom, Alsthom-Atlantique of France, and Interatom of West Germany. Each of these vendors developed small factory-fabricated nuclear plants of nominally 125 MWe. In addition, the Soviet Union focused on a more traditional commercial plant of 440 MWe. Several of the Soviet 440 MWe plants were ultimately constructed, primarily in Eastern Bloc countries. None of the small prefabricated plants were ever ordered, due in part to the high uncertainty in the cost of these new and unproven systems.

The International Atomic Energy Agency (IAEA) was established in 1957 to facilitate the deployment of civilian nuclear power throughout the world, especially for developing countries. The first evidence that I could find of their direct support of smaller-sized nuclear plants was a conference held in September of 1960 at IAEA headquarters in Vienna, Austria.[21] The conference, which included over 250 attendees from 40 countries, focused on small and medium power reactors because of their better suitability to lesser developed countries. The conference demonstrated great interest by both suppliers and buyers; however, very few of the small reactor concepts ever materialized.

More than 20 years later, the IAEA launched the Small and Medium Power Reactor Project Initiation Study in 1983 to again survey the availability of smaller-sized nuclear power plants and the level of interest by developing countries to deploy these plants.[22] This study involved multiple surveys issued to reactor vendors and emerging countries that had expressed interest in initiating nuclear power programs. Interestingly, the global nuclear landscape had changed dramatically in the intervening 20 years since the 1960 study. At the time of the earlier study, reactor vendors had a sufficient number of orders for large plants and had no interest in developing small units for what was perceived to be a minor and uncertain export market. By 1983, however, domestic markets for these vendors appeared quite uncertain, and they began to seriously consider the export market in new emerging countries. A technical meeting was held at the IAEA at the conclusion of their 1983 study, at which 23 small- and medium-sized power reactor designs were presented, representing 17 different vendors in 9 different countries. Table 2.2 summarizes the reactor designs included in the 1983 IAEA study.

Table 2.2 **Summary of small and medium power reactor concepts included in the 1983 IAEA project initiation study[22]**

Country	Vendor	Concept	Type
Canada	Atomic Energy of Canada, Ltd	CANDU 300	PHWR
France	Framatome/Technicatome	NP 300	PWR
Germany (F.R.)	Kraftwerk Union	PHWR 300	PHWR
	BBC/HRB	HTR 100/300/500	HTGR
Italy	Ansaldo/Nira	PWR 272	PWR
	Ansaldo/Nira	CIRENE 300	HWLWR
Japan	Hitachi	BWR 500	BWR
	Toshiba	BWR 200/300/500	BWR
	Mitsubishi	PWR 300	PWR
Sweden	ASEA/ATOM	PIUS 500	PWR
UK	Rolls Royce	PWR 300	PWR
	GEC	Magnox	GCR
	National Nuclear Corporation	Magnox 300	GCR
US	General Electric	Small BWR	BWR
	Babcock & Wilcox	CNSS	PWR
	Babcock & Wilcox	CNSG	PWR
	General Electric	HTGR	HTGR
	General Electric	MRP	LMR
USSR	Atomenergoexport	VVER 440	PWR

Unifying characteristics of the small-sized designs reviewed in the IAEA study were as follows: (1) an emphasis on a shortened and more predictable construction schedule, (2) utilization of proven systems and components to enhance confidence in the new designs, (3) a high level of factory prefabrication, and (4) recognition of diverse siting considerations in developing countries. These qualities remain as cornerstones of most SMRs being developed today. From the buyer's perspective, economic competitiveness was important but was not the only decision-making factor. Other important factors included total project cost, safety, flexibility of use, and infrastructure requirements—quite similar to modern considerations, as discussed in Chapter 8.

In response to the 1983–1985 IAEA study on small and medium reactors, the Nuclear Energy Agency (NEA) assembled an expert group in 1991 to also evaluate this class of designs.[23] As mentioned earlier, the IAEA was created to facilitate the expansion of nuclear power to newcomer countries and therefore is focused largely on the needs and interests of developing countries. In contrast, the NEA, which is contained within the Organization for Economic Cooperation and Development (OECD), is largely focused on the interests of those countries that already have well-established economies. The IAEA and the NEA are complementary and often cooperate, but their different missions and orientations are frequently reflected in the scope of their studies and the tone of their conclusions. This difference is especially apparent in the two agencies' approaches to evaluating smaller-sized nuclear reactors.

The first thing you will notice is that the NEA study was initiated 6 years after the completion of the IAEA study. This reflects a more conservative reaction by developed countries to the global interest in smaller-sized reactors. The established reactor vendors in the OECD countries had made their fortune from selling large plants and were slow to admit that this market had substantially vaporized. Second, the NEA devoted a significantly larger fraction of its study to analyzing the economic factors associated with small and medium reactors and attempted to predict market opportunities. In contrast, the IAEA study focused more on the technical features of the small reactor concepts in relation to the energy and infrastructure conditions of their member countries. Finally, the conclusions of the NEA study were cautiously encouraging about the market opportunity for smaller-sized reactors but only if they are able to be competitive with large nuclear plants. This completely misses the point that SMRs are first and foremost intended for markets that cannot afford or utilize large plants. This is a reoccurring difference that plays out again during the resurgence of SMRs post-2000.

The NEA expert group provided a thorough analysis of the motivations and challenges for deploying smaller-sized commercial power reactors. Some of the key benefits that they articulated are as follows:

- potential to open up additional energy markets;
- providing a valuable contribution to CO_2 reduction;
- better response to slow growth rates of energy demand;
- better fit for small electricity grids; and
- good match for the replacement of older, smaller fossil fuel plants.

All of these benefits remain valid today, as will be discussed in later chapters. The expert group identified several challenges for small- and medium-sized reactor deployment, including their first-of-a-kind nature, the multitude and diversity of concepts, economies of scale that favor large plants, and regulatory uncertainties. These challenges also remain valid today and will be discussed in more detail in Chapter 9. The overriding challenge for their deployment potential at the time of the NEA study was the prevailing negative public attitude toward nuclear power. While this may not be as dramatic today, it is a factor that must always be carefully considered and managed.

The study reviewed 17 small- and medium-sized reactor concepts that were intended for electricity generation or cogeneration of electricity and heat as well as seven concepts intended for process heat only. These concepts and their developers are shown in Figure 2.4. You will notice that some of the concepts are the same as were included in the IAEA study, while some were dropped and other new designs had emerged. Several of the designs evolved from the US-led Advanced Light-Water Reactor and Advanced Liquid-Metal Reactor programs.

A point of clarification is useful here. While this book is focused on small *modular* reactors, that is, reactors that have power capacities that are nominally below 300 MWe and are predominantly prefabricated in a factory, both the IAEA and the NEA prefer to group small reactors (<300 MWe) and *medium* reactors (300–700 MWe) together without regard to their design or construction characteristics. Unfortunately and confusingly, both nomenclatures share the same acronym: SMR. Throughout this book, I will use "SMR" to denote only "small modular reactor" and avoid the acronym entirely when discussing IAEA and NEA studies.

I commented earlier on several of the designs shown in Figure 2.4, some of which evolved out of the US advanced reactor programs. One design that was relatively new at the time of the NEA study was the Safe Integral Reactor (SIR), which deserves special note. Frequently a new reactor design is derived from a precursor design, sometimes an entire lineage of designs. In the case of SIR, it was derived from the Minimum Attention Plant (MAP) concept, which in turn was derived from very early central station plant and maritime reactor designs developed by Combustion Engineering.[24] The MAP concept incorporated numerous innovations, including an integral configuration of the primary system with self-pressurization, elimination of primary coolant pumps, replacement of soluble boron in the primary coolant with solid burnable absorbers in the core, and the elimination of control rods. In the conference paper by Turk and Matzie that describes the 150–300 MWe MAP concept, they list the key elements of their design philosophy:

> …maximize the reliance on existing LWR technology, simplify the design by eliminating systems and components, build in inherent safety by use of passive mechanisms, and maximize advantages released by small size.[24]

They went on to list several specific design goals, such as reduced core power density, significantly greater volume of primary coolant per megawatt of power, and increased operator response time. I highlight these features because the exact same elements are being embraced by many of the current SMR designers. So it should not be surprising that the overall appearance of the MAP reactor concept and its successor, SIR, bears a striking resemblance to contemporary designs.

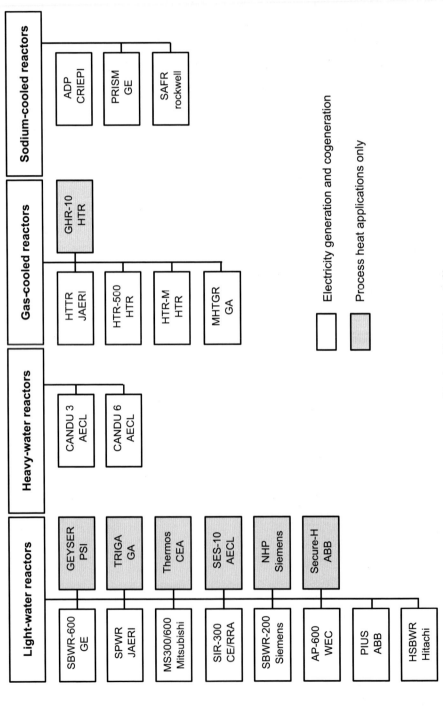

Figure 2.4 Small- and medium-sized reactor concepts reviewed in the 1991 NEA study.[23]

The SIR design was developed by a consortium of the UK Atomic Energy Agency, Combustion Engineering, Rolls Royce, and Stone and Webster. They started by replacing some of the more radical features of MAP with more conventional PWR features, including standard control rods, standard fuel elements, and the addition of coolant pumps. But importantly, they retained their "state-of-the-art approach to safety."[25] The SIR consortium matured the concept through the preliminary design, including the three independent passive safety systems for decay heat removal.[26] The design was proposed as a candidate for the US Advanced Light-Water Reactor program; however, it was not selected for funding by the US Department of Energy, which selected instead the Westinghouse AP-600 design and the General Electric SBWR design. This left the SIR consortium short on financial resources to complete the design. However, some of its features surfaced a decade later in a new SMR design developed by Westinghouse, which acquired the nuclear portion of Combustion Engineering in 2000.

It should be apparent from this extensive history lesson on SMRs that the notion of them as a fad is completely dispelled, at least as a one-time fad. Perhaps they are instead a reoccurring fad, similar to the Teenage Mutant Ninja Turtles or the hula hoop, destined to reappear every decade or so but never to be sustained. In the next chapter, we will take a closer look at the more recent history of SMRs—the period from roughly 2000 to today—in the hope of assessing the staying power of their current popularity.

References

1. Loewen EP. *The USS seawolf sodium-cooled reactor submarine*. Washington, DC: Address to the American Nuclear Society Local Section; May 17, 2012.
2. Aircraft nuclear propulsion, GlobalSecurity.org. Available at: www.globalsecurity.org/wmd/systems/anp.html.
3. Muckenthaler FJ. *The tower shielding facility—its glorious past*. Oak Ridge National Laboratory; 1993. ORNL-12339.
4. United States Air Force, defenseimagery.mil photograph no. DF-SC-83-09332.
5. *Report to congress: review of manned aircraft nuclear propulsion program*. Comptroller General of the United States; 1963.
6. Frenkel KA. *Resuscitating the atomic airplane: flying on a wing and an isotope*. Scientific American; December 5, 2008.
7. Suid LH. *The Army's nuclear power program: the evolution of a support agency*. Greenwood Press; 1990.
8. Energy Information Administration (www.eia.doe.gov) and Nuclear Energy Institute (www.nei.org), August 2008.
9. Nuclear News. *Am Nucl Soc* January 1968;**11**(1):38.
10. Anderson TD, et al. *An assessment of industrial energy options based on coal and nuclear systems*. Oak Ridge National Laboratory; July 1975. ORNL-4995.
11. Ingersoll D, Houghton Z, Bromm R, Desportes C, McKellar M, Boardman R. Extending nuclear energy to non-electrical applications. In: *Proceedings of the 19th Pacific Basin nuclear conference*. Canada: Vancouver, B.C.; August 24–28, 2014.

12. Spiewak I, Klepper OH, Fuller LC. *Technical and economic studies of small reactors for supply of electricity and steam*. Oak Ridge National Laboratory; February 1977. ORNL/TM-5794.

13. Klepper OH, Smith WR. Studies of a small PWR for onsite industrial power. In: *Proceedings of the American power conference 39th annual meeting*. Chicago, IL; April 1977.

14. Bupp IC, Derian JC. *Light water: how the nuclear dream dissolved*. New York: Basic Books; 1978.

15. Kemeny J. *Report of the President's commission on the accident at three mile island*. Washington, DC: US Government Printing Office; 1979.

16. Martel L, Minnick L, Levey S. *Summary of discussions with utilities and resulting conclusions*. Electric Power Research Institute; 1982. EPRI-RP-1585.

17. Weinberg AM, Spiewak I, Barkenbus JN, Livingston RS, Phung DL. *The second nuclear era*. Praeger Publishers; 1985.

18. Forsberg CW, Reich WJ. *Worldwide advanced nuclear power reactors with passive and inherent safety: what, why, how and who*. Oak Ridge National Laboratory; September 1991. ORNL/TM-11907.

19. US Nuclear Regulatory Commission. Available at: www.nrc.gov/reactors/new-reactors/col.html; September 2014.

20. Heising-Goodman CD. Supply of appropriate nuclear technology for the developing world: small power reactors for electricity generation. *Appl Energy* 1981;**8**:19–49.

21. *Proceedings from the conference on small and medium power reactors*. International Atomic Energy Agency; September 5–9, 1960.

22. *Small and medium power reactors: project initiation study Phase 1*. International Atomic Energy Agency; 1985. IAEA-TECDOC-347.

23. *Small and medium reactors: status and prospects*, vol. 1. Nuclear Energy Agency; 1991.

24. Turk RS, Matzie RA. The minimum attention plant: inherent safety through LWR simplification. In: *Proceedings of the winter meeting of the American Society of Mechanical Engineers*. Anaheim, CA; December 7–12, 1986.

25. Dettmer R. *Safe integral reactor*. Institute of Electrical Engineers, IEE Review; November 1989.

26. Matzie RA, Longo J, Bradbury RB, Tear KR, Hayns MR. Design of the safe integral reactor. *Nucl Eng Des* 1992;**136**:73–83.

The rise of current small modular reactors (2000–2015)

3

In the previous chapter, I reviewed the ebb and tide of nuclear energy in the US prior to the year 2000, culminating in a redirection of the nuclear industry toward smaller, more robust reactor designs. With no new plants ordered in nearly three decades, the industry, at least the vendor and manufacturing sectors of it, had atrophied. Those that survived did so by supporting the existing fleet of plants with fuel reloads or maintenance services. Also, the nuclear power research and development (R&D) community had largely vaporized. The R&D budget for the US Department of Energy's (DOE) Office of Nuclear Energy dwindled steadily throughout the 1990s and bottomed out at zero in 1998. Although it looked bleak for the US nuclear industry, factors were already playing to again reverse its fortune and provide at least the opportunity for a comeback. These factors began to come into focus around the turn of the century. As "a rising tide raises all boats," this resurgence of interest in nuclear power also reinvigorated interest in small modular reactors (SMRs). Consequently, many new SMR designs emerged after 2000, some of which are still progressing toward the market place.

3.1 Precursors to the nuclear renaissance

By 2000, it was apparent that the US was in the early phase of the "second nuclear era," as Weinberg called it, or more commonly referred to as the nuclear renaissance. This dramatic change from just a few years earlier was enabled by a number of circumstances that evolved during the preceding decade, including the following:

- a continually growing demand for electricity and a steadily dwindling margin of generating capacity;
- the new awareness of the impact of energy supply on national and energy security;
- a growing concern for the environmental impacts of large-scale burning of fossil fuels, especially its impacts on global climate change; and
- the excellent safety and performance record of the existing fleet of water-cooled reactors.

So before discussing the actual renaissance, a brief digression back into the 1990s is instructive.

In order to maintain stability in the electricity distribution grid and ensure that demand is met, the government requires utilities to maintain an excess generating capacity, referred to as the capacity margin or reserve margin. The size of the margin is both regionally and seasonally dependent, and targets are set based on projected demand. Averaged over the nation, the summer demand is consistently higher than the winter demand, so the summer load becomes the most constraining period for total capacity

Small Modular Reactors. http://dx.doi.org/10.1016/B978-0-08-100252-0.00003-3

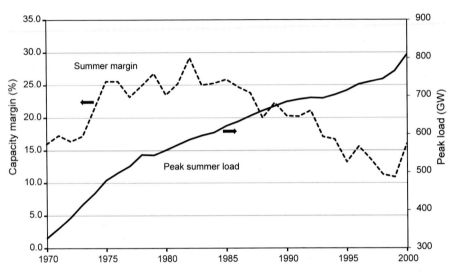

Figure 3.1 Summer capacity margin and peak summer load in the contiguous 48 states.[1]

and capacity margin. Figure 3.1 shows the US (excluding Alaska and Hawaii) capacity margin and peak summer load over the period from 1970 to 2000.[1] It is evident from the figure that the aggressive build-out of nuclear plants in the 1970s helped to create a significant capacity margin—well above target levels of nominally 15%. By 1995, however, the steady increase in demand and no new capacity addition resulted in a critically low margin. So it was clear that a new generating capacity had to be built—but what?

Adding to the compelling need for additional generating capacity was the increasing awareness of the true cost of energy imports. The instabilities in the Middle East that resulted in Iraq's invasion of Kuwait in 1990 and a subsequent major US military engagement there highlighted the vulnerabilities and obligations resulting from the US' thirst for imported oil. This created a new urgency for the development and expansion of domestic energy sources. The term "energy security" became popular during this time in close alignment with national security.

New generating capacity from renewables, especially biofuels and wind, was already on the rise, but renewables alone could not account for the hundreds of gigawatts that would be needed. The price of natural gas was still high and volatile, causing some utilities to walk away from installed gas-fired plants due to their poor economics. The capital cost of modern coal plants was on par with nuclear plants and rising, plus coal plants carried the added uncertainty of potential economic penalties for their carbon emissions. As discussed in Chapter 1, the debate on global climate change was just cranking up during this time, which served to cast doubt on the economics of building more coal or natural gas plants due to potential carbon emission restrictions. The door was open for nuclear power. But was nuclear a credible option given the dismal performance of the industry during the 1970s and 1980s?

Fortunately, in the intervening years since the 1970s, the track record of the nuclear industry has improved remarkably. Without the rush to build new plants, the industry

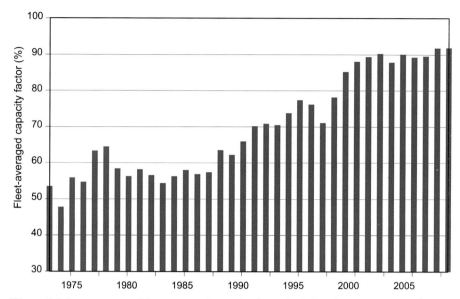

Figure 3.2 Improvement of fleet-averaged capacity factor for US nuclear power plants.[2]

turned to a methodical refining of the operations of the existing nuclear fleet. Figure 3.2 shows the progress in the fleet's performance as measured by the fleet-averaged capacity factor from 1973 through 2008.[2] When factored over the nominally 100 nuclear units comprising the fleet, the improvement in capacity factor from 50% to 90% during this period added the equivalent of greater than 20 GW-class plants to the US generating capacity. At the same time, a combination of refined regulatory oversight and transparent sharing of operational experience across the industry produced an impeccable safety record. Given the need for new generating capacity and the challenges or concerns with fossil fuel options, the stage was set for a dramatic nuclear comeback. In short, the nuclear industry was well positioned to "save the day," be it with large or small reactors or both.

3.2 Restarting the nuclear industry

As the US crossed over into the new millennium with increasing interest in nuclear power, the surviving reactor vendors quickly dusted off their most current designs. Westinghouse, who had developed their AP-600 design under the DOE's Advanced Light-Water Reactor program, chose to upsize the design to the AP-1000 with a power capacity of 1140 MWe. Even though the AP-600 had just received design certification approval from the US Nuclear Regulatory Commission (NRC) in 1999, a new design certification application had to be submitted and reviewed. The application was submitted to the NRC in 2002 and was eventually granted several years later after multiple amendments. The other domestic mid-sized design that had been certified during the Advanced Light-Water Reactor program was General Electric's SBWR-600. General Electric chose to upsize that design to the 1500 MWe ESBWR. The ESBWR was submitted to the NRC for design certification review in 2005, which was eventually granted in 2014. A number of foreign reactor vendors also jumped into the US market by seeking NRC design certification. These included Toshiba's ABWR (1350 MWe), Areva's EPR (1650 MWe), and Mitsubishi's USAPWR (1700 MWe). The NRC, which was primarily engaged in overseeing the existing fleet of operating reactors and reviewing power uprate requests, suddenly found its inbox quite full. In 2006, they created a new division, the Office of New Reactors, to focus on new applications for design certification.

To help kick-start the process of reviving the nuclear industry, in 2002 the DOE initiated the Nuclear Power 2010 (NP-2010) program, which targeted the construction of a new nuclear plant in the US by 2010. The main thrust of the program was to fully exercise the new licensing framework that the NRC had developed over the preceding two decades. The original 10 CFR Part 50 framework required the plant owner to secure a license to construct a new plant and then secure a second license to operate it after completion of the construction. The troublesome part of this process was the fact that public hearings were required for both licenses. It did not take the antinuclear community very long to figure out that they could easily bankrupt a hopeful owner by stalling the operating license process, thereby depriving the owner of revenue from his or her shiny (and expensive) new plant. The new restructured licensing framework, 10 CFR Part 52, allows an owner to secure a combined construction and operating license (COL) from a single application with the only public hearing occurring prior to construction. The new framework further allows a potential owner to qualify a site separately and in advance of the COL through the early site permit (ESP) process. The main purpose of the NP-2010 program was to exercise this new licensing framework by cost-sharing with industry the preparation of the ESP and COL applications. The program was highly successful and generated more than 20 new ESP or COL applications to the NRC, ultimately facilitating the order and construction of the first new builds in the US in over 30 years—2 AP-1000 units at the Vogtle site in Georgia and 2 AP-1000 units at the VC Summer site in South Carolina.

3.3   Restarting the nuclear R&D community

In the world of smaller-sized nuclear reactors, a few traditional vendors started developing SMR designs based on their past experiences with some of the earlier designs that had emerged during the 1980s and 1990s. But many new designs evolved from a fresh look at nuclear power by the research community, which was also spinning up rapidly with a new infusion of DOE funding. After being denied any federal budget for nuclear energy R&D in 1998, the DOE succeeded in getting appropriated funds in 1999 for the Nuclear Energy Research Initiative (NERI). The NERI program not only breathed new life into the research community but also represented a major departure from the previous approach to federal R&D. In particular, nuclear research in the days of the Atomic Energy Commission was largely determined and directed by Commission staff in Washington, DC. In contrast, the NERI program was basically an open call to the remnants of the R&D community to submit new ideas for research. This was the equivalent of offering free beer at a college party—the DOE was flooded with proposals. The solicitation also encouraged collaborations among universities, national laboratories, and industry to further re-energize the community. This too was quite successful.

3.3.1 The nuclear energy research initiative

One of the winning consortia that received a 3-year grant in the 1999 NERI solicitation was led by Westinghouse and included the University of California–Berkeley and the Massachusetts Institute of Technology. The consortium also included nonfunded contributions from the Politecnico di Milano in Italy and the Japan Atomic Power Company. They proposed a new SMR concept, which they called the International Reactor Innovative and Secure (IRIS). The initial technical goals for the IRIS concept were chosen to be consistent with the target goals of the NERI solicitation, which emphasized designs that were safer, smaller, faster, cheaper, and more secure.[3] The IRIS goals were extremely ambitious, including such goals as a modular arrangement with module capacity between 50 and 200MWe, 15-year core lifetime, capacity factor greater than 95%, no maintenance, no access to vessel internals for the projected 60-year plant lifetime, and all of this with a total cost of electricity to not exceed a bargain price of 4¢/kWh.

The IRIS design was based on mature water-cooled reactor technology and utilized an integral primary system configuration much like the earlier Consolidated Nuclear Steam Generator (CNSG), Minimum Attention Plant, and Safe Integral Reactor designs discussed in the previous chapter. After the conclusion of the 3-year DOE-sponsored research project, Westinghouse's Chief Scientist, Mario Carelli, continued to expand the international team, eventually achieving an unprecedented 20 organizations from 10 countries, including the US, Italy, Japan, Brazil, United Kingdom, Croatia, Lithuania, Spain, Mexico, and Estonia. Not all of the charter goals for the design survived the engineering rationalization process, but the one goal that remained during the life of the project was safety.[4]

Although prelicensing discussions of IRIS were initiated with the US NRC in 2005, the activity was suspended 2 years later and eventually terminated when Westinghouse withdrew from the IRIS Consortium in 2010. Other Consortium members, led by the Politecnico di Milano and Tokyo Institute of Technology, continued to advance the IRIS design for a few years. However, without an identified nuclear vendor to tie together the universities, laboratories, and suppliers within the Consortium, all commercialization effort was eventually stopped.

A personal note: After hearing Carelli present a description of the IRIS project during a student colloquium at the University of Tennessee in 2000, I recommended that Oak Ridge National Laboratory (ORNL) join the IRIS Consortium, which it did in 2001. Serving as the ORNL lead, I coordinated our contributions in the areas of reactor physics, shielding, probabilistic risk analysis, and instrumentation and controls. Participating in the Consortium was a career-changing experience that both solidified my enthusiasm for SMRs and greatly expanded my knowledge of reactor design principles. Most importantly, I learned that safety is a design choice and a goal against which all design decisions must be benchmarked. Carelli formalized this in what he called "safety by design," which became the mantra of the IRIS Consortium. Also, my engagement in the IRIS project was enormously rewarding from the perspective of the broad international relationships that it fostered. Carelli managed to assemble a talented team of researchers and engineers who worked seamlessly across national borders and cultural differences. It has been several years since I left the IRIS Consortium, but I still consider many of my IRIS colleagues to be close professional friends.

The original IRIS concept represented the light-water version of a broader class of SMRs called STAR, which stood for Secure, Transportable, and Autonomous Reactor. The intent of this class of concepts was to provide a safe and secure energy option for developing countries. Several other STAR concepts were pursued, primarily by national laboratories funded through subsequent NERI grants or within the other major R&D program at that time, the Generation IV program. Most of the STAR concepts used more exotic technologies such as lead or lead-bismuth as the primary coolant to achieve very long core lifetimes. Some of the different STAR concepts that resulted were the STAR-LM (for "liquid metal"), STAR-H2 (focused on hydrogen production), SSTAR (for "Small" STAR), and SuperSTAR. The designs differed according to the particular application and the professional opinions of the concept leaders. IRIS was the only STAR concept that jumped from the research community to a commercial environment before eventually being abandoned. Of the liquid-metal-cooled concepts, the 20 MWe SSTAR concept had the most sustained development effort within the national laboratories.[5]

A second consortium that had proposed a new water-cooled SMR concept in the inaugural NERI solicitation in 1999 was also awarded funding. This team included researchers from the Idaho Engineering and Environmental Laboratory, Oregon State University, and Nexant, a subsidiary of Bechtel. This team sought to develop the Multi-Application Small Light Water Reactor (MASLWR) concept with top-level goals of a simple design with maximum use of passive safety features, a construction schedule of less than 2 years, maximum use of shop fabrication, unrestricted transport of the units by rail or roadways, and, like IRIS, a cost of electricity less than 4¢/kWh.[6] Demonstration of the concept's safety performance was also a key goal.

The initial MASLWR concept was a 1000 MWt design with four horizontal U-tube steam generators. This quickly proved to be overly complex and uneconomical, and the module design was simplified to a small 150 MWt (35 MWe) integral primary system configuration using an internal helical-coil stream generator. In order to produce an output similar to conventional large plants, 30 reactor modules and associated turbine/generator equipment were assembled in a single plant, producing a total output of 1050 MWe. In addition to developing the innovative SMR concept and multimodule plant design, the MASLWR project team constructed an electrically heated, scaled simulator facility at Oregon State University to demonstrate the safety performance of the concept. After the 3-year NERI project was completed, an Oregon State team lead by Jose Reyes used the simulator to further develop the MASLWR concept. Several changes and refinements were made to the design to improve its performance and commercial viability. In 2007, a new company, NuScale Power, was formed for the purpose of commercializing the resulting design. I will address the NuScale design more thoroughly in the next section as one of the current US SMR designs that is still progressing toward licensing and deployment.

3.3.2   The Generation IV program

The Generation IV program was established in 2000 by the DOE to study advanced reactor technologies. This quickly evolved into an international program called the Generation IV International Forum (GIF) and included 10 charter countries: Argentina, Brazil, Canada, France, Japan, South Africa, South Korea, Switzerland, the United Kingdom, and the US. Whereas the NERI program focused on re-engaging the research community, the Generation IV program focused on the broader international community and sought to achieve a consensus on commercially viable advanced reactor technologies. A large cohort of managers was assembled into a complex layering of committees that would bring a tear of joy to the eye of any bureaucrat. There were concept-level committees, technology-level committees, evaluation committees, screening committees, integration committees, and of course the obligatory advisory committees. Despite the bureaucratic challenges, the collective set of committees succeeded in evaluating an initial compilation of nearly 100 concepts and assimilated them into a few promising options. The final set of six concepts, actually six classes of concepts, that were selected for further study included the Very High Temperature Reactor, the Sodium-Cooled Fast Reactor, the Lead-Cooled Fast Reactor, the Gas-Cooled Fast Reactor, the Supercritical Water Reactor, and the Molten Salt Reactor.[7] Specific reactor designs from each of these concept classes continue to be developed by various countries.

The Generation IV program goals of improved safety, economics, sustainability, and proliferation resistance were "size neutral," and specific applications of smaller-sized reactors were considered. But as the program was implemented through the GIF, which included mostly individuals and organizations from traditional nuclear programs, the six advanced reactor concepts evolved quickly to larger-sized designs. The program acknowledged interest by several of the GIF countries in small, integral water-cooled designs but did not include them in the program because water-cooled technology was felt to be sufficiently mature. This put design teams like the IRIS Consortium and the MASLWR team at a disadvantage because their designs were too mature to be part of the Generation IV program but were considered to be too innovative to be part of the Nuclear Power-2010 program, which was focused on accelerating the deployment of previously certified designs. So it was up to the partnering organizations to fund further development—a tough challenge given the large cost of such an endeavor. This situation changed in 2006 with the initiation of a new DOE program, the Global Nuclear Energy Partnership (GNEP).

Of little note was a report by the DOE issued in 2001 in response to a Congressional request to characterize the benefits of SMRs, especially for powering remote sites in Alaska and Hawaii.[8] The study, which focused on SMR concepts less than 50 MWe, concluded that there were no significant technical challenges in deploying any of several candidate designs reviewed, and that their economic competitiveness in remote regions of the two states appeared to be promising.

3.3.3 The Global Nuclear Energy Partnership program

Beginning in 2005, a small group of government and laboratory leaders met privately to develop the scope of a new, broad program to promote the international expansion of nuclear energy, especially for emerging countries. Initially labeled the Global Nuclear Energy Initiative, the draft program was rolled out to a broader community of researchers in the fall of 2005. I was among those from ORNL who attended the rollout meeting and was excited to discover that the program included an element focused on SMRs. By the time the program was funded by Congress in 2006, it had been relabeled the GNEP and involved the participation of the US, France, Japan, and the United Kingdom. By 2008, more than 20 countries were signatories of the GNEP Principles with a similar number of countries observing. The key elements of GNEP included expanding the use of nuclear power, minimizing nuclear waste, enhancing nuclear safeguards, developing appropriately sized reactors, developing recycling technologies, and establishing reliable fuel services. Whether by original intent or a seizure of opportunity, the US portion of the GNEP program became focused on developing technologies for advanced fuel reprocessing and recycling.

Due to the overwhelming emphasis of GNEP on fuel cycle technologies, the small reactor element had a minor role and received less than 0.5% of the roughly $150M annual GNEP budget. I was selected by the DOE to be the technical lead for this program element because of my involvement in the IRIS project and because of my incessant pleading for the job. Although the level of importance of the small reactor component within GNEP was disappointing, it provided the first funding since the initial NERI grants explicitly targeting small reactors. The IRIS design became a central focus of GNEP's small reactor element, called the Grid-Appropriate Reactor campaign, because it was the only SMR design in 2006 that had significant commercial involvement. Also, the international basis of the IRIS Consortium and the extensive publication record of the Consortium made the design well recognized in many of the countries involved in GNEP.

The original intent of including a small reactor component in GNEP was to develop commercially viable reactor designs that would be suitable for emerging countries, which strongly implied that they should be smaller and more robust than the megaplants currently on the market. Because of the export focus of the program and the perception that small reactors were of no interest for domestic deployment, the US Congress zeroed the Grid-Appropriate Reactor funds in the 2008 federal budget. Actually, many in Congress did not like the GNEP program in general and canceled it a year or two later. But as anyone who works with federal programs knows, programs do not actually disappear; they just resurface under different names. In the case of GNEP, it was split three ways. The bulk of the R&D activities emerged as the well-funded Advanced Fuel Cycle program, and the international relationships portion of GNEP was relabeled the International Framework for Nuclear Energy Cooperation. The Grid-Appropriate Reactors portion of GNEP became the stand-alone SMR program, although without any appropriated funding. The brief run of the Grid-Appropriate Reactors campaign and the proposed follow-on SMR program served to re-expose the broader nuclear community to the merits of smaller-sized reactors and catalyzed a surprising amount of utility interest domestically. The programs also helped to stimulate the emergence of several new SMR designs, many of which are still being actively developed.

3.4 Renewed interest by the military

The renewed national interest in nuclear power and the DOE's attention to SMRs generated a new look at nuclear power by the US Army and Air Force. The Air Force was the first to step forward in 2007 with a cautious interest. Fortunately, their interest was not to power military aircraft, as they had done in the 1950s, but rather to provide secure, dedicated power to their domestic bases. They issued a formal Request for Information in early 2008 to solicit interest from potential vendors, owners, and operators of small nuclear power units to provide electricity and potentially process heat for liquid fuel production—a major component of the Air Force's total energy consumption.[9] A few months later, the US Army also advertised interest in using small nuclear plants to power their domestic bases. But unlike the original Army Nuclear Power Program discussed in the previous chapter, they now favored an approach similar to the Air Force, whereby they would offer to lease federal land and provide a long-term power purchase agreement to a commercial consortium that would design, build, and operate the plant for the Army.

The sudden interests of both the Air Force and the Army to provide secure power to domestic bases quickly prompted a number of new small and very small reactor concepts to emerge. One concept proposed to the Air Force for this application was the Global Energy Module (GEM-50), developed by Babcock & Wilcox under the technical leadership of Jeff Halfinger. The 50 MWe GEM-50 was an integral pressurized water reactor design and had several features in common with Babcock & Wilcox's earlier SMR designs, including the propulsion unit for the Otto Hahn merchant ship and the CNSG and Consolidated Nuclear Steam Supply designs for industrial heat applications. Although Babcock & Wilcox had been one of the major US large-plant vendors in the 1970s, they had dropped out of this market and focused on supporting the US Navy. Their effort in developing the GEM-50 concept helped to set the stage for their introduction in 2009 of a new SMR design, designated mPower, which is discussed more in the next section.

In an effort to address the similar interests of the Air Force and the Army for nuclear power to support their domestic bases, a multiagency group was formed in late 2009, reporting to the Office of the Secretary of Defense. The committee consisted of representatives from the Air Force, Army, Navy, DOE, and NRC. I participated with the DOE on this committee and got to see firsthand both the enthusiasm and the reluctance of the Department of Defense (DOD) to re-embrace nuclear power. I also got to experience just how effective it is to kill a good idea by assigning it to a multiagency committee. Progress was painfully slow. Fortunately, the US Congress included language in the 2010 DOD funding authorization bill prompting DOD to conduct a study on the feasibility and merits of deploying nuclear power units on domestic bases.[10] The joint committee contracted the Center for Naval Analysis to conduct the study, which culminated in a final report published in early 2011.[11] The study concluded that SMRs were the best option due to the relatively low power demand of most installations. It also concluded that SMRs could contribute to electric energy assurance for critical military facilities while also addressing federal mandates for the reduction of greenhouse gas emissions. The study further observed that the new SMR designs offer promising solutions; however, their licensing and economics were yet unproven. The DOD continues to be cautiously interested in the utilization of SMRs at their installations but seems content to let the commercial industry take the lead in bringing the new designs to market.

3.5   Emergence of contemporary SMR designs in the US

A detailed plan was drafted for the proposed DOE SMR program in 2009. It was initially modeled after the short-lived Grid-Appropriate Reactor program plan and was refined with broad input from the nuclear community. It included two key elements: (1) a cost-shared government-industry collaboration to accelerate the design and licensing of near-term SMRs and (2) a cross-cutting R&D program to further develop technologies supporting both near-term and especially longer-term advanced SMR concepts. The proposed SMR program appeared to be widely popular, but as a new federal program, it could not start until Congress passed a federal budget for the country with the program explicitly included. Unfortunately, Congress had degenerated into decision gridlock and could not agree on a federal budget in 2010 and again in 2011, leaving the federal government to operate under a Continuing Resolution. Although the SMR program remained unfunded during this time, it appeared to be quite popular politically and helped to motivate several new SMR concepts and designs to emerge in both the commercial and research communities.

As mentioned earlier, the IRIS design was the most visible US SMR until the mid-2000s. However, the original MASLWR concept that had been developed as an early NERI project was quietly being matured by Oregon State University and went commercial in 2007 with the formation of a new start-up company: NuScale Power, LLC. NuScale's cofounders were Jose Reyes, chair of the Oregon State University Nuclear Engineering Department, and Paul Lorenzini, retired president of Pacific Power & Light. They quickly assembled a small cadre of industry experts to advance the NuScale design and officially opened a preapplication project with the US NRC the following year.[12] Although the NuScale SMR design incorporates several improvements relative to the MASLWR concept, it maintains many of MASLWR's innovative features, including a small, simplified module that can be operated in a shared reactor pool with many identical units to provide a very robust and scalable plant design.[13] The NuScale design gained immediate traction with the DOE and DOD because its small power size (50 MWe per module) was a good fit for the power requirements of several federal installations.

Like most new start-up companies, NuScale had its share of investment challenges and nearly failed in early 2011. Later that year, Fluor Corporation stepped in as the major investor and strategic partner, putting NuScale on a firm financial basis. In 2013, several additional companies, such as Rolls-Royce, partnered with NuScale to provide support for the design and manufacture of plant components. Later in 2013, NuScale was selected by the DOE to receive a major grant worth over $200M as part of the DOE's SMR program, which had finally been funded by Congress in fiscal year 2012.

A personal note: After working for nearly 35 years at ORNL, I left my position there to follow my passion for SMRs. I became convinced that it was up to industry to make SMRs a reality, and I was fortunate to be able to join NuScale Power. This opportunity was an ideal fit for me because of NuScale's corporate culture of innovation and a driving focus on design safety. Similar to Carelli's constant demand for "safety by design," Reyes and his team were implementing the same principle in the NuScale design to an even greater extent. To me, the design appeared to have a truly remarkable level of resilience and safety—the kind of characteristics that Weinberg was convinced should be the future of nuclear power. I fully subscribe to this belief.

Funding for the DOE SMR program was slow to materialize despite broad political support. In fact, its popularity was so apparent by 2011 that it became a political bartering chip by various political individuals and organizations. The DOE persisted in arguing the value of the SMR program and was able to work through several political challenges. In one of those little ironies of life, Congress finally passed a federal budget specifying funding for the SMR program on the very day of my departure from ORNL in December 2011.

Shortly after the emergence of NuScale's SMR design in 2007, Babcock & Wilcox announced their entry into the commercial SMR market in 2009. Responsibility for the design was subsequently transferred to Generation mPower, LLC, which is a partnership of Babcock & Wilcox and Bechtel. The mPower design is an integral pressurized water reactor and builds on Babcock & Wilcox's extensive historical experience in both large and small reactor designs and current experience in naval propulsion systems. The design introduced several changes from the GEM-50 design that Babcock & Wilcox had proposed a year earlier to the Air Force, such as forced circulation of the primary coolant and the use of conventional straight-tube steam generators. The reference mPower plant contains two 180 MWe reactor modules with independent turbine/generator systems.[14]

The mPower developers initiated a preapplication licensing review with the NRC in 2009. In 2012, they were the first SMR design to be selected for a major grant to complete design certification and deployment as part of the DOE's newly funded SMR program. Although the design work progressed very rapidly between 2009 and 2013, Babcock & Wilcox announced in early 2014 that they would significantly reduce the pace of development due to investment challenges. This emphasizes perhaps the largest hurdle in bringing a new reactor design to market—the substantial and protracted commitment of investment funds. I will revisit this challenge in a later chapter.

One year after Babcock & Wilcox introduced the mPower design, Holtec International introduced their entry into the SMR competition: the 140 MWe Small Modular Underground Reactor (HI-SMUR). Although not a traditional reactor vendor, Holtec is recognized in the nuclear industry for their design and manufacturing of nuclear fuel racks. They assembled a diverse team led by the newly created subsidiary, SMR, LLC, to develop the HI-SMUR design. Contributors to the design included the Shaw Group and the United Kingdom National Nuclear Laboratory. Roughly a year later, the design was upgraded to 160 MWe and relabeled the SMR-160. The SMR-160 design is not a true integral reactor like NuScale or mPower but is referred to as a "compact loop" design to acknowledge that the external steam generator vessel is flanged directly to the reactor vessel, thus eliminating large external piping.

The fourth serious contender in competition for the near-term DOE grants was Westinghouse. Although they had led the IRIS Consortium since 1999, they withdrew from it in 2010. In early 2011 they introduced a new SMR design that utilized a number of components and features from their highly successful AP-1000 design while maintaining an integral primary system configuration similar to IRIS.[15] The 800 MWt (225 MWe) module is intended to be deployed as a single unit or as a two-module plant. In 2012, they initiated prelicensing discussions with the NRC; however, Westinghouse announced in late 2013 that they were significantly reducing their efforts on their SMR design in order to concentrate on their growing AP-1000 business line.

Another smaller-sized nuclear plant was being developed during this time as part of the DOE's Next Generation Nuclear Plant (NGNP) program.[16] The NGNP was different from the four SMR designs discussed above for several reasons: (1) it was not a specific design but rather an envelope of specifications, (2) it focused on high-temperature process heat applications rather than electricity generation, and (3) it required very different fuel, materials, and coolants than conventional water-cooled reactors. The NGNP was based on the Very High Temperature Reactor concept, which was one of the six advanced technology concepts selected by the Generation IV program in 2002. Although originally planned to produce an outlet temperature of 1000°C for the primary helium coolant, the NGNP design goal was subsequently lowered to 850°C due to significant material challenges. The high outlet temperature is desired to support several process heat applications such as the production of hydrogen using efficient thermo-chemical processes, advanced oil recovery from shale and tar sands, and steam reforming of natural gas. The anticipated power generation capability of a single NGNP unit was 250–300 MWe.

Through much of the NGNP program, which was terminated in 2012, three design options were developed in parallel by three industry teams lead by Westinghouse, General Atomics, and Areva. The Westinghouse design was the Pebble Bed Modular Reactor (PBMR) design that had been developed by Eskom in South Africa. The General Atomics design was the Modular High-Temperature Reactor (MHR), which was fundamentally the same design as was developed in the early 1980s. The Areva

New Technology Advanced Reactor for Energy Supply design used a prismatic block arrangement for the moderator similar to the MHR but used an indirect gas and steam cycle power conversion system.

With the termination of the NGNP program, the future of the three candidate designs is in question, although all remain viable for future deployment if sufficient market interest materializes. Similarly, the Power Reactor Inherently Safe Module (PRISM) design, which was developed originally in the 1980s, continues to be maintained by General Electric for potential deployment where the advantages of a fast-spectrum reactor are desired, that is, for improved fuel utilization (fuel breeding) or for waste management (actinide transmutation).

In addition to the nonwater-cooled reactor designs offered by these traditional reactor vendors, several new start-up companies formed between 2000 and 2010 in the hopes of leveraging the federal funding that was being offered in the GNEP and SMR programs. The new start-up that received the most visibility, largely due to their aggressive marketing style, was Hyperion Power Generation. The company was formed in 2007 to commercialize a 25 MWe "nuclear battery" using a very exotic concept developed at Los Alamos National Laboratory. The design was later changed to a more conventional design, although its lead-bismuth coolant technology is still quite advanced compared to water-cooled reactor technology. The company leadership also changed, as did the name, which is now Gen4 Energy. The concept, which is intended for remote and largely unattended operation, is several years away from potential licensing and deployment but has received funding from the R&D portion of the DOE SMR program to help further develop the technology.

In 2010, another start-up company was formed, Advanced Reactor Concepts, LLC, to commercialize a small sodium-cooled fast-spectrum reactor.[17] The ARC-100 concept was developed substantially by former DOE national laboratory researchers and was a 100 MWe sodium-cooled reactor similar in concept to PRISM. The focus of the ARC-100 development effort was to provide a very robust and secure nuclear power plant design to meet the emerging energy demands of developing countries. The company was not successful in acquiring sufficient investment funding and is no longer pursuing active development of the design.

In the world of gas-cooled reactors, General Atomics introduced their newest line of small helium-cooled reactors in 2009: the Energy Multiplier Module (EM2). Unlike their thermal-spectrum MHTGR and MHR designs, the EM2 is a small fast-spectrum reactor intended to extend fresh fuel resources and manage nuclear waste through the use of a breed-and-burn fuel cycle. Considerable R&D is required to fully develop the fuel form and materials, for which they are cost-sharing research with the DOE. Table 3.1 summarizes several of the commercial SMR designs that have emerged in the US since 2000.

Multiple other new SMR concepts emerged during this time period, spurred by the community's enthusiasm for SMRs, the hope of federal dollars, or both. Some have succeeded in receiving R&D funding from the DOE SMR program, and others have not. A few may see the eventual fruits of their labor, but history suggests that most will not.

Table 3.1 Summary of current US commercial SMR designs

SMR design	Designer	Configuration	Coolant circulation	Electrical output (MWe)
Water-cooled				
mPower	Generation mPower	Integral	Forced	180
NuScale	NuScale Power	Integral	Natural	50
SMR-160	SMR	Compact loop	Natural	160
W-SMR	Westinghouse	Integral	Forced	225
Gas-cooled				
EM2	General Atomics	Loop	Forced	265
MHR	General Atomics	Loop	Forced	280
Liquid-metal-cooled				
G4M	Gen4 Energy	Pool	Natural	25
PRISM	General Electric	Pool	Forced	311

3.6 Slowing of the nuclear renaissance

In contrast to the roaring enthusiasm and bright outlook for nuclear power in 2000, it became apparent by roughly 2007 that implementation of the nuclear renaissance was going to take a lot longer than thought but hopefully less than the 400 years that spanned the European cultural renaissance. Although there was an initial flood of license and certification applications submitted to the NRC, some were beginning to be suspended, and very few actual orders were occurring. Instead, there were delays in the certification of the new AP-1000 and ESBWR designs and very slow progress in the implementation of federal loan guarantees and other risk-mitigating features that had been approved as part of the Energy Policy Act of 2005. The situation got worse in 2008 when a global financial crisis occurred, wreaking havoc on US and international economies. The crisis had two chilling effects on new builds. First, it made investment money for large expensive nuclear plants very risky and difficult to secure. Second, the crisis also generally reduced the demand for electricity as many consumers struggled to pay bills. Adding insult to injury was the precipitous drop in the cost of natural gas due to the new gas-recovery process called fracking.

Another major blow to the nuclear industry occurred on March 11, 2011, when a major earthquake off the coast of Japan created a devastating tsunami that destroyed four operating reactor units at Japan's Fukushima Daiichi plant. Three of the four units suffered major core damage in the ensuing days, requiring a large-scale evacuation of people in the surrounding region.[18] The basic cause of the destruction was that the height of the tsunami exceeded the protective barriers, causing a loss of off-site power to the plant and flooding the emergency diesel generators. The emergency backup batteries were able to provide critical power for only several hours. Without power to operate the reactor coolant pumps, the reactor cores overheated and melted.

In reaction to this accident, Japan closed all of its nuclear power stations and seemed certain to walk away from nuclear power entirely. Germany decided to phase out all of their nuclear power, and Italy, who was on the brink of re-engaging nuclear energy, slammed the door closed again. Everyone in the nuclear industry held their breath in anticipation of more countries following suit. Fortunately, they did not.

Somewhat surprising to me was that many of the factors that put the skids on the resurgence of nuclear power in the US resulted in a bolstering effect for SMRs. As examples, the financial crisis preferentially penalized large capital-intensive projects, such as gigawatt-class nuclear power plants, and the appreciable drop in electrical demand also made utility executives less inclined to order very large generating capacities. In contrast, smaller-sized plants with lower up-front costs and the ability to add new capacity in smaller increments were immediately more appealing, even to the larger utilities. In a relatively short few years we observed that the industry shifted from where there was almost no domestic customer interest in SMRs to a point that entire conferences were being held on the topic.

Even the accident at the Fukushima Daiichi plant helped to cement interest in SMRs. I was still working at ORNL at the time of the Fukushima accident, and the ORNL public relations office directed many inquiries regarding the accident to me. Fielding calls from journalists was a relatively new experience for me and one that I generally dreaded. However, this particular round of calls changed my generally negative opinion of journalists. To my surprise, nearly all of them had done their homework and came prepared with at least a basic understanding of a nuclear plant design. Also, they had no preconceived agenda or story—they just wanted to better understand what had happened in Japan and its implications. Several of them had read articles about SMRs and offered their speculation that SMRs might be a better answer than the design of the plants at Fukushima. Despite my enthusiasm for what SMRs can offer with respect to plant safety and resilience, my response to these journalists was always carefully measured. The reality is that many existing plants around the world, and all of the plants operating in the US, would have demonstrated a more resilient response than the particular designs of the Fukushima Daiichi units. Furthermore, the new large plant designs that utilize passive safety systems such as AP-1000 offer an additional level of resilience to such extreme accidents. Still, the early response of those journalists based on limited research was encouraging to me and proved to be the foretelling of a similar response by the broader nuclear community.

3.7   International SMR activities

The preceding sections of this chapter are focused on the US experience with SMRs, as is the bulk of this book. However, SMRs have been and continue to be a global phenomenon. In total, there are more than 20 SMR designs with significant commercial investments worldwide and easily twice that number if one includes concepts being developed by major research organizations. In several respects, some countries have outpaced the US in the development of new SMR designs, including licensing and construction. For example, the System Integrated Modular Advanced Reactor (SMART) design developed in the Republic of Korea received Standard Design Approval from the Korean Nuclear Safety and Security Commission in 2012. The Russian Federation has licensed their design of a barge-mounted SMR, the KLT-40S, and began construction of the first two units in 2012. Argentina announced in early 2014 that construction had been initiated for the Central Argentina de Elementos Modulares (CAREM) prototype unit.

As discussed in the last chapter, both the International Atomic Energy Agency (IAEA) and the Nuclear Energy Agency (NEA) studied emerging trends and opportunities for small- and medium-sized reactors in the 1980s and early 1990s. The IAEA continued an active engagement with the global community during the 2000s, driven largely by the personal enthusiasm of Vladimir Kuznetsov, who was assigned to the IAEA in 2003. Kuznetsov initiated or contributed to several small- and medium-sized reactor studies, beginning with a technical meeting in 2004 to review the status of current development and deployment interests.[19] It was the first IAEA meeting in roughly a decade focused on small- and medium-sized reactors, which had grown in number from less than 20 commercial designs in 1995 to more than 30 designs in 2004. The 2004 technical meeting catalyzed a broader effort by the IAEA to collect design information from both the commercial and research communities and publish a massive compendium of small and medium reactor designs. Descriptions of the concepts and designs were released in two separate reports: one pertaining to designs with conventional refueling[20] and one pertaining to designs that precluded on-site refueling.[21] Collectively, the 2 reports totaled more than 1600 pages and described more than 60 designs.

About the time that the US initiated the Generation IV program, the Russian Federation formed a similar program operating within the IAEA. The International Project on Innovative Nuclear Reactors and Fuel Cycles (INPRO) was established to help ensure that nuclear energy can contribute to meeting the energy needs of the twenty-first century in a sustainable manner. Although the goals of the two programs were quite similar, the Generation IV program focused on developing six specific reactor technologies while the INPRO program focused on developing methodologies for evaluating advanced technologies. In 2007, INPRO initiated a project called Common User Considerations for the purpose of defining common characteristics needed by potential users of new plants in developing countries.[22] Although originally intended to explore the suitability of small and medium reactors for developing countries, the project struggled to stay focused on that mission. One challenge was that developing countries had little or no awareness of evolving designs for small- and medium-sized reactors. In addition, their nuclear programs were insufficiently developed to meaningfully discuss requirements and acceptance criteria at a detailed level. I also observed that there was not unanimous support for smaller-sized reactors among the INPRO team, which limited the opportunity for a meaningful outcome from the project.

The Common User Considerations project was my first direct involvement in the IAEA, and it was a real eye-opener. Foremost, it gave me a much better understanding of the desperate energy situation in many countries throughout the world. Nuclear energy provides a highly enticing option for them but also a huge dilemma. The financial constraints and physical infrastructure in their countries make a compelling case for the use of smaller-sized power plants; however, their desire for very low project risk appears to preclude the new SMR designs that are highly promising but not yet operational (or even licensed). Despite the many challenges of the project, it helped to initiate an important dialogue between the developers of new small reactor designs and a compelling group of potential customers: the emerging countries of the world. When I attended a subsequent INPRO meeting on small- and medium-sized reactors five years later in 2013, it was striking how much the exchange between SMR developers and potential users had progressed and how much more sophisticated the understanding was of the emerging countries to the benefits and challenges of SMRs.

In parallel to the INPRO Common User Considerations project, the IAEA conducted a 2-year project beginning in 2008 looking at the economic competitiveness of small- and medium-sized reactors.[23] I also participated in this project, and it was my first serious exposure to economists, a community that I had been largely sheltered from during my career. Although somewhat painful, due to the very foreign terminology and thought processes, it proved to be highly valuable in establishing a more quantitative basis for articulating why smaller-sized plants could be competitive with large plants. The approach, which I found quite appealing, was not to attempt to predict what the cost of a particular plant or class of plants would cost but rather to understand factor-by-factor how a smaller-sized reactor can incrementally offset the traditional economy-of-scale factor. This issue is so central to the viability of SMRs that I will devote an entire chapter to it later. The final report of the project on economic competitiveness unfortunately got caught in internal delays and was not published until 2013.

In 2010, Hadid Subki replaced Kuznetsov as the leader of the small- and medium-sized reactor program within the IAEA Division of Nuclear Power. Since that time, Subki has carried forward an aggressive program of meetings on various aspects of technology, licensing, and deployment of small- and medium-sized reactors. He also initiated the routine publishing of a compendium of major designs, the most recent of which was released in September of 2014.[24] This compendium lists 22 water-cooled designs and 9 gas-cooled designs (liquid-metal-cooled designs were moved to a separate report). Keep in mind, though, that the IAEA typically groups small reactors with medium-sized reactor designs (capacity between 300 and 700 MWe), so some of these designs are larger than what I would consider to be an SMR. Table 3.2 lists my assessment of current commercial SMRs, which are each briefly described in Chapter 2 of the *Handbook of Small Modular Nuclear Reactors*, also published in September of 2014.[25]

The other major international nuclear organization, the NEA, remained relatively silent on SMRs during the 2000s. In fact, there was no significant attention on SMRs at the NEA since their study in 1991, which was discussed in the previous chapter. The silence was broken in 2011 when they issued a report providing their surmise of the status of small- and medium-sized reactors.[26] The study reviewed the spectrum of current and emerging designs but focused on the economic viability of small- and medium-sized reactors. Although providing a lot of encouraging data on the potential markets, the report was skeptical of the ability of the smaller-sized plants to be competitive in large plant markets. As in the 1991 NEA study, it is my opinion that the 2011 study again misses the point that SMRs primarily target nontraditional markets that are not well served by large plants.

Table 3.2 **Summary of leading commercial SMR designs being developed globally**[24,25]

Country	SMR	Designer	Configuration	Output (MWe)	Modules per plant
Light-water-cooled					
Argentina	CAREM	CNEA	Integral	27	1
China	ACP-100	CNNC	Integral	100	Up to 8
China	CNP-300	CNNC	Loop	300–340	1
France	Flexblue	DCNS	Loop	160	1
S. Korea	SMART	KAERI	Integral	100	1
Russia	ABV-6M	OKBM	Integral	8.5	2
Russia	KLT-40S	OKBM	Compact loop	35	2
Russia	RITM-200	OKBM	Integral	50	1
Russia	VBER-300	OKBM	Compact loop	300	1
US	mPower	Generation mPower	Integral	180	2
US	NuScale	NuScale Power	Integral	45	Up to 12
US	SMR-160	SMR	Compact loop	160	1
US	W-SMR	Westinghouse	Integral	225	1
Heavy-water-cooled					
India	PHWR-220	NPCIL	Loop	235	2
India	AHWR-300-LEU	BARC	Loop	304	1
Gas-cooled					
China	HTR-PM	INET	Pebble bed	105	2
S. Africa	PBMR	PBMR	Pebble bed	100	2
US	GT-MHR	General Atomics	Prismatic	150	1
US	EM2	General Atomics	Prismatic	265	2
Liquid-metal-cooled					
Japan	4S	Toshiba	Sodium	10 or 50	1
Russia	SVBR-100	AKME	Lead-bismuth	101	1
US	PRISM	General Electric	Sodium	311	2

From the past two chapters, you can see that there has been a persistent interest in SMRs in the US and globally. Despite this sustained interest, SMRs have yet to become a significant force in the commercial nuclear power industry. The level of SMR-related activities worldwide, and especially in the US, has grown rapidly since roughly 2000—I cannot possibly mention all of them here. To better understand why interest in SMRs has recently escalated rapidly and to address the question of whether or not they will ever become a part of the nuclear power future, I provide in Part Two, Fundamentals and features, a discussion of the major attributes and benefits associated with SMRs. In Part Three, Promise to reality, I offer the perspectives of several different groups of customers on the value proposition of SMRs. I also address some of the challenges and hurdles that must be resolved if SMRs are to become a reality.

References

1. *Statistical abstract of the United States: 2012.* US Census Bureau; 2012.
2. *Monthly energy review.* Energy Information Administration; March 2009.
3. *IRIS development and objectives.* Westinghouse Electric Company; September 9, 1999. IRIS-W-02 Rev 0.
4. Carelli MD, et al. The design and safety features of IRIS. *Nucl Eng Des* 2004;**230**:151–67.
5. Smith CF, et al. SSTAR: the US lead-cooled fast reactor (LFR). *J Nucl Mater* 2008;**376**:255–9.
6. Modro SM, et al. *Multi-application small light water reactor final report.* Idaho National Engineering and Environmental Laboratory; December 2003. INEEL/EXT-04–01626.
7. *A technology roadmap for generation IV nuclear energy systems.* Generation IV International Forum; December 2002. GIF-002–00.
8. *Report to congress on small modular nuclear reactors.* US Department of Energy; May 2011.
9. *Request for information.* US Air Force; January 17, 2008. AFRPA-08-R-0005, posted on Federal Business Opportunities. www.fbo.gov.
10. *National defense authorization act for fiscal year 2010.* 2010. HR 2647, Report 111–288, Section 2845.
11. King M, Huntzinger L, Nguyen T. *Feasibility of nuclear power on US military installations.* Center for Naval Analyses; March 2011. CRM D0023932.A5/REV.
12. Memorandum from T.J. Kenyon to W.D. Reckley. *Summary of pre-application kickoff meeting with NuScale power, Inc. on the NuScale reactor design and proposed licensing activities.* US Nuclear Regulatory Commission; August 7, 2008.
13. Reyes Jr JN. NuScale plant safety in response to extreme events. *Nucl Tech* May 2012;**128**:153–63.
14. Halfinger JA, Haggerty MD. The B&W mPower scalable, practical nuclear reactor design. *Nucl Tech* May 2012;**128**:164–9.
15. Memmott JJ, Harkness AW, Wyk JV. Westinghouse small modular reactor nuclear steam supply system design. In: *Proceedings of the international conference on advanced power plants.* Chicago, IL; June 24–28, 2012.
16. *Next generation nuclear plant pre-conceptual design report.* Idaho National Laboratory; November 2007. INL/EXT-07–12967, Rev. 1.
17. Wade D. ARC-100: a sustainable, modular nuclear plant for emerging markets. In: *Proceedings of the international conference on advanced power plants.* San Diego, CA; June 2010.
18. *Fukushima Daiichi: ANS committee report.* American Nuclear Society; March 2012.
19. *Innovative small and medium sized reactors: design features, safety approaches and R&D trends.* International Atomic Energy Agency; May 2005. IAEA-TECDOC-1451.
20. *Status of innovative small and medium sized reactor designs 2005: reactors with conventional refueling schemes.* International Atomic Energy Agency; March 2006. IAEA-TECDOC-1485.

21. *Status of innovative small and medium sized reactor designs without on-site refueling.* International Atomic Energy Agency; January 2007. IAEA-TECDOC-1536.

22. *Common user considerations (CUC) by developing countries for future nuclear energy systems: report of state 1.* International Atomic Energy Agency; 2009. NP-T-2.1.

23. *Approaches for assessing the economic competitiveness of small and medium-sized reactor.* International Atomic Energy Agency; 2013. NP-T-3.7.

24. *Advances in small modular reactor technology developments.* International Atomic Energy Agency, Supplement to the IAEA Advanced Reactors Information System (ARIS); September 2014.

25. Ingersoll DT. Small modular reactors for producing nuclear energy: international developments [Chapter 2]. In: *Handbook of small modular nuclear reactors.* Cambridge, UK: Woodhead Publishing; 2014.

26. *Current status, technical feasibility, and economics of small nuclear reactors.* Nuclear Energy Agency; June 2011.

Part Two

Fundamentals and features

Nuclear power 101: understanding nuclear reactors

The first chapter of this book focused on the importance of energy and justifications for nuclear power as an important part of our clean energy future. In Chapters 2 and 3, I reviewed the history of nuclear power in general and of small modular reactors (SMRs) in particular. However, I have not yet described specific features of SMRs in any significant detail. Admiral Hyman Rickover has been quoted as saying, "The Devil's in the detail, but so is salvation." You should find some of both in the next several chapters as I begin to focus on characteristics that are common to many SMR designs and some that are unique to specific designs. But before diving into the details of SMRs, it may be helpful for readers to review the basic features, functions, and technologies of commercial nuclear power plants. For those readers who are already familiar with reactor basics, you can skip this chapter without any loss of continuity. But for those less comfortable with the subject, this brief overview will help to explain references to certain reactor technologies and terms in subsequent chapters.

4.1   Basic power plant features and functions

The primary function of a commercial power plant is to produce heat. In that regard, a nuclear plant is functionally the same as a coal plant or a natural gas plant. The only functional difference in the three types of plants is how the heat is produced. In the case of coal and natural gas plants, the heat is produced by burning coal or gas in a furnace, thus releasing chemical energy in the form of direct heat. In the case of a nuclear plant, the heat is produced by splitting atoms within the fuel, which releases a large amount of nuclear energy. Although the nuclear reactor system is quite different from a coal furnace or a natural gas furnace, the rest of the power plant hardware, referred to as the "balance of plant," looks quite similar in all three types of power plants and has the same basic function of converting the generated heat into electricity.

Figure 4.1 shows a simplified diagram of a notional nuclear power plant. The nuclear fuel, which is usually in a ceramic or metal form, is encased in a protective cladding and consolidated into discrete assemblies that comprise the reactor core. All nuclear processes occur within the reactor core, which is contained within a very stout reactor vessel. Control rods, which contain a special material that absorbs neutrons, move in and out of the core to turn the fission reaction off and on. The control rod drive mechanisms needed to move the control rods are typically located outside of the reactor vessel but may be inside the vessel in some integral configurations. A coolant, referred to as the primary coolant, is circulated through the core to transfer

Small Modular Reactors. http://dx.doi.org/10.1016/B978-0-08-100252-0.00004-5

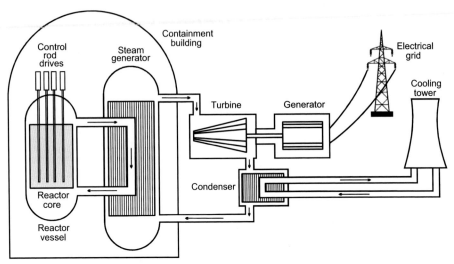

Figure 4.1 Notional diagram of a nuclear power plant.

the heat from the core to a steam generator. In the steam generator, which is simply a densely packed bundle of metal tubes, heat is transferred from the primary coolant to a secondary coolant, typically water. The secondary coolant boils to produce steam that is injected into a turbine, causing it to rotate rapidly. The rotation of the turbine shaft, which is connected to a generator, converts the rotational energy into electricity through a process called electromagnetic induction, and the electricity goes to the power grid. The steam that exits the turbine is cooled in a condenser, which is basically a steam generator running in reverse to convert the steam back into water. The condensed water is then recirculated back through the main steam generator. The cooling function of the condenser is provided by a separate loop of circulating water that is cooled externally to the power plant using either a nearby water source, such as a river, lake, or ocean, or the atmosphere using a cooling tower.

An important material found within some reactor cores is the neutron moderator. Some nuclear fuels such as uranium work more efficiently if the neutrons that create fission have slow velocities, referred to as thermal neutrons. The function of the moderator is to take the very fast-moving neutrons that are emitted during the fission process and slow them down to thermal neutron energies. Therefore, reactors that use uranium fuel are referred to as thermal reactors and will always include moderator material in the reactor core. The moderator material may be the same material as the primary coolant, which is the case for water-cooled reactors. Graphite is also a common moderator material, especially for gas-cooled reactors (GCRs). An alternative approach is to use a different nuclear fuel, such as plutonium, which fissions efficiently using fast-moving neutrons and does not require a neutron moderator. Not surprisingly, these types of reactors are referred to as fast reactors. Since water is quite efficient at slowing down neutrons, fast reactors cannot use water as the primary coolant. Differences in coolant and moderator materials are the primary distinctions among reactor types and will be discussed further below.

Specific nuclear plant designs may include additional heat transfer loops to further isolate the primary coolant from the turbine/generator equipment or may eliminate the steam generator by using the primary coolant to drive the turbine directly. Also, the steam generator may be outside of the reactor vessel (loop-type design) or inside the reactor vessel (integral-type design). It is important to note, however, that the external coolant loop used to cool the condenser is always separate from the primary coolant in order to isolate the reactor coolant system from the environment. Another plant feature that provides isolation is the containment structure that surrounds the reactor system.

The existing US fleet of nominally 100 nuclear power plants all use water as the primary coolant. They also address only one of the US' energy needs: centralized base-load electricity generation. It is possible to extend the use of nuclear energy to other energy demands, such as distributed electricity generation and industrial process heat applications; however, these applications may be better served by a different plant design and underlying technology than the existing fleet of large water-cooled reactors. For example, using a gas or liquid-salt coolant can enable the reactor system to operate at much higher temperatures than water-cooled reactors, which is advantageous for some industrial processes. Likewise, a liquid-metal-cooled reactor changes the nuclear fission dynamics in ways that allow the reactor to either breed new fuel faster than it consumes fuel or reduce the amount of radiological hazard in the spent fuel. So even though there is tremendous operational experience with water-cooled reactors, many countries are also pursuing the development and deployment of more advanced reactor technologies.

4.2 Reactor generations

The terminology reactor "generation" emerged in the late 1990s to help clarify distinctions between nuclear plants that were currently operating and new plant designs being developed using more advanced technologies. While the terminology has been modestly useful, the boundaries between generations are a bit fuzzy, and the basis for assigning a generation to new emerging designs is unclear. Fundamentally, Generation I is intended to include the early prototype reactors that were designed and built to gain familiarity with the various reactor technologies. These early plants had relatively low power outputs (less than 200 MWe) and provided the engineering basis for the current fleet of large commercial plants, which are categorized as Generation II. During the US hiatus of new plant orders in the 1980s and 1990s, several new plant designs were developed, designated as Generation III. These designs incorporated lessons learned from the previous generation of plants, especially regarding design simplification, standardization, and increased use of passive safety features. However, no Generation III plants were ordered in the US even though two designs, the AP-600 and the ABWR, were certified by the US Nuclear Regulatory Commission (NRC). Globally, Generation III-type designs have been built in Japan, the Republic of Korea, and China.

In 2000, the US Department of Energy initiated a program to leapfrog the unordered Generation III designs by developing a new generation of designs, designated Generation IV, based on advanced technologies. The Generation IV program resulted in identifying six broad classes of advanced reactors that showed potential to dramatically improve the performance of previous generations, especially with regard to safety, economics, sustainability, and proliferation resistance.[1] Moreover, several of the advanced concepts were expected to enable the extension of nuclear energy to more than base-load electricity production, particularly process heat applications and waste management functions. The six specific classes of advanced concepts that were evaluated and selected include the following:

- Very High-Temperature Reactor (VHTR);
- Super-Critical Water-Cooled Reactor (SCWR);
- Molten Salt Reactor (MSR);
- Sodium-Cooled Fast Reactor (SFR);
- Lead-Cooled Fast Reactor (LFR); and
- Gas-Cooled Fast Reactor (GFR).

These six classes of reactor systems can be grouped into two fundamental categories: (1) high-temperature reactors that are principally for use with process heat applications (VHTR, SCWR, MSR, and GCR) and (2) fast-spectrum reactors that are principally for use with fuel cycle applications (SFR, LFR, and GCR). All six reactor types can be used to produce electricity with higher conversion efficiencies than traditional light-water reactors (LWRs) due to their higher operating temperatures; however, their real strength is in addressing nonelectrical energy demands.

About the same time that the Generation IV program was initiated, a resurgence of interest by US utilities to build new nuclear capacity spurred some vendors of Generation III designs to produce updated versions of their designs. These upgraded Generation III designs are referred to as Generation III+. Two examples of Generation III+ designs are the Westinghouse Advanced Passive (AP-1000) and the General Electric Economical and Simplified Boiling Water Reactor (ESBWR). The first four AP-1000 plants are under construction in China. Another four units are under construction in the US: two units at the Vogtle site in Georgia and two units at the VC Summer site in South Carolina. No ESBWRs have been ordered yet, but several utilities in the US are considering the design for future construction.

The emergence of new light-water-cooled SMR designs in the mid-2000s created a bit of a dilemma for the industry in terms of generation terminology. Some people classify them as Generation III+, although integral SMRs are significantly different in design configuration than the other Generation III+ designs. On the other hand, they do not qualify as Generation IV since they use traditional water-cooled reactor technology. When asked my opinion on this issue, I sometimes playfully assign integral water-based SMRs as being Generation 3.825.

4.3 Reactor technology classes

Although not used consistently, the term "technology" in the context of nuclear reactor types most often refers to the coolant used to remove heat from the reactor core. The choice of coolant typically influences the overall design and substantially impacts the choice of other materials in the reactor system, such as the fuel,

fuel cladding, core structures, and reactor vessel.[2] The most common reactor coolants are the following:

- water, including light (H_2O) and heavy (D_2O) water;
- gas, including carbon dioxide and helium;
- metal, including sodium, lead, or lead-bismuth alloy; and
- fluoride salt, including options with solid or dissolved fuel.

The particular characteristics, including benefits and disadvantages of each coolant type, are summarized in the following sections.

SMRs are not a specific reactor technology since small reactor designs have been developed using each of these different coolants. Instead, SMRs are simply a design option within a selected reactor technology, that is a choice by the designer to limit the output capacity of the reactor. This choice may or may not impact the arrangement of the plant components, but the fundamental technology of the design remains mostly a function of the choice of coolant. A more technical discussion of SMR technologies, including physical properties of the various coolants, is provided by Neil Todreas in the opening chapter of the *Handbook of Small Modular Nuclear Reactors*.[3] What should become obvious in the discussion below is that there is truly no "silver bullet" and that all technologies have both advantages and disadvantages.

4.3.1 Water-cooled reactors

At the end of 2014, there were 435 commercial power reactors in operation worldwide.[4] Of these, 419 are water-cooled reactors, 15 are gas-cooled, and 1 is metal-cooled, as reflected in Figure 4.2. Given that 96% of all commercial reactors worldwide are water-cooled, using this technology for a new reactor design is an overwhelming advantage in terms of engineering, regulatory, and operational experience. Supply chain availability is also a clear advantage. From a technology perspective, water is a highly familiar and available commodity and completely benign in terms of industrial handling. The primary disadvantage of water for reactor cooling

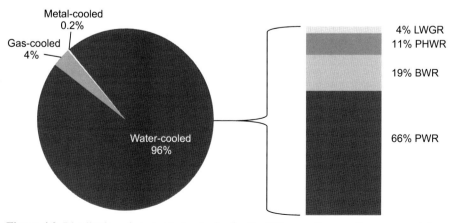

Figure 4.2 Distribution of reactor technologies for the 435 currently operating commercial power reactors worldwide as of December 31, 2014.[4]

is that it has a low boiling temperature (100°C at atmospheric pressure). In order to be economically viable, the plant must be able to achieve a reasonably high efficiency in converting heat to electricity, which implies operating the reactor at temperatures above 300°C. This temperature can be achieved using water as the primary coolant but requires that the reactor system be operated at a relatively high pressure. The high system pressure requires thicker (more expensive) pressure vessels and increases the energetics of system leaks.

Variations within the class of water-cooled reactors include pressurized water reactors (PWR), boiling water reactors (BWR), pressurized heavy-water reactors (PHWR), and light-water-cooled, graphite-moderated reactors (LWGR). Figure 4.2 shows the relative distribution of these reactor types within the 419 water-cooled commercial reactors. These designs are typically large loop-type plants, that is the primary coolant circulates from the reactor vessel into one or more external steam generator vessels and back into the reactor vessel. A common design option for water-cooled SMRs is the integral primary system configuration in which the steam generators are moved inside the reactor vessel to avoid the external loops. Most integral SMRs are also PWRs and sometimes referred to as iPWRs.

Loop-type PWRs and BWRs have been used for commercial power generation for over 50 years worldwide. Collectively, they represent 85% of all commercial reactors globally and 100% of commercial reactors in the US. The key difference between the two types is where the water boils. In a PWR, the primary water is maintained at a very high pressure (150 times normal atmospheric pressure) in order to keep the water from boiling within the reactor vessel. The primary coolant is circulated through the reactor vessel and transfers the heat from the reactor core to an external heat exchanger where the hot primary water causes lower pressure water on the secondary side of the steam generator to boil and produce steam. Commercial BWRs operate with the primary coolant at roughly half the pressure of a PWR, which allows the water to boil within the reactor core. The steam produced from the boiling primary water is sent directly to the turbine, thus eliminating the need for a separate steam generator. The primary coolant also serves as the neutron moderator for both the PWR and the BWR. More detailed descriptions of BWR and PWR reactors are given in Chapters 3 and 4 of the *Reactor Concepts Manual* maintained by the NRC.[5]

The PHWR reactor is a different type of PWR that uses heavy water instead of light water as the primary coolant and also as the neutron moderator. In heavy water, the hydrogen in normal (light) water molecules is replaced with deuterium, producing a form of water that is chemically identical to normal water but with improved nuclear properties. Because heavy water does not absorb as many neutrons as light water, natural uranium may be used as a fuel, consequently avoiding a uranium enrichment infrastructure and cost. The reactor core is made up of an array of individually pressurized coolant and fuel channels that allow refueling of the reactor while it continues to operate. This is in contrast to PWRs and BWRs that must be fully shut down during refueling. The LWGR is similar to a BWR because the water coolant boils within the core region; however, it uses graphite as the neutron moderator. Pressurized coolant and fuel channels within the LWGR allow for continuous refueling similar to PHWRs.

4.3.2 Gas-cooled reactors

GCRs offer potential operational and safety benefits over water-cooled reactors. A major operational motivation for considering this technology is improved energy conversion efficiency afforded by a higher reactor operating temperature. For instance, water-cooled reactors have a practical maximum temperature limit of about 350°C, which yields a heat-to-electricity conversion efficiency of roughly 32–34%. By comparison, a GCR can operate at temperatures up to 800–850°C and yield an energy conversion efficiency exceeding 40% when using conventional steam turbine equipment or as high as 50% when using more advanced gas turbine equipment.

From a safety perspective, GCRs typically use a lower core power density and a high heat capacity core, which helps limit fuel temperatures following a loss-of-coolant accident. Because they use a different fuel form and cladding, they avoid steam/zirconium cladding chemical reactions that can release explosive hydrogen gas under accident conditions in LWRs. Unlike traditional PWRs and BWRs, some GCR designs have the capability of being refueled during full-power operation, which provides some operational convenience and higher plant availability.

Several GCRs have been built and operated using either CO_2 or helium to cool the reactor core. Commercial GCRs use a graphite neutron moderator, which absorbs fewer neutrons than water-moderated reactors. The first-generation GCRs were built in the United Kingdom and France and used natural uranium metal fuel and magnesium or magnesium alloys for the cladding. Subsequent plants used low-enriched uranium–oxide fuel clad with stainless steel. All 15 of the GCRs operating as of the end of 2014 are located in the United Kingdom. Beginning in the late 1950s in Germany, a new generation of helium-cooled GCRs was developed that used a very robust graphite-coated particle fuel. The coated fuel particles are formed into fuel compacts that are surrounded by a graphite moderator, which typically takes the form of hexagonal blocks (prismatic variant) or spherical balls (pebble-bed variant).

The US constructed and operated a prototype GCR at the Fort St. Vrain plant from 1979 to 1989. It used prismatic block, graphite fuel elements, a uranium–thorium fuel cycle, and helium coolant. Currently the Japanese are operating a 30-MWt prismatic helium-cooled high-temperature test reactor. Three pebble-bed reactors have been constructed and operated: the German AVR and THTR test reactors and the currently operating Chinese HTR-10 test reactor. South Africa built upon the German experience in developing their Pebble Bed Modular Reactor design; unfortunately, financial hardships stalled their efforts to construct a commercial plant. The Chinese, based on their experience with HTR-10, are currently in the process of constructing a pair of 100-MWe pebble-bed reactors in a commercial-sized plant designated the HTR-PM.

4.3.3 Metal-cooled reactors

Development of metal-cooled reactors began shortly after the initial development of water-cooled reactors. Liquid metals have several attractive features for reactor applications, including high thermal conductivity, high boiling points, and relatively low melting points. These features offer potential benefits such as high core power density, small core volumes, and thin-walled pressure vessels compared to LWRs. Candidate metals that have been studied for commercial nuclear power include sodium, lead, and lead-bismuth. These coolants have low neutron moderation properties; therefore, the core contains a high fraction of fast neutrons, which yield higher numbers of neutrons per fission than thermal neutrons. These excess neutrons can be used for a variety of purposes such as producing new fuel from fertile material; in fact, they can produce more fuel than they consume. A potential reduction in high-level waste repository requirements is also an attractive benefit of these reactors. The fast neutron spectrum in the core is favorable in converting plutonium and some other transuranic actinides into short-lived fission products to reduce or eliminate long-lived, high-heat-producing radioactive waste in the spent nuclear fuel.

Even though SFRs are one of the Generation IV concepts, there is considerable global experience with SFRs. The US, France, Japan, Russia, and the United Kingdom have all built test, prototype, or demonstration SFRs. The only commercial SFR is in Russia. China and India are currently constructing demonstration SFRs. To date, the operation of SFRs has proved to be challenging, due substantially to industrial issues associated with the handling and use of sodium metal, which has an energetic reaction with water. Additional design features must be added to avoid sodium leakage, which creates additional challenges for the design and economic competitiveness of SFRs.

Lead and lead-bismuth coolants have also been considered for fast reactors. They have a number of advantages relative to sodium such as higher boiling temperatures, low coolant void reactivity feedback, and low chemical reactivity with air, water, and steam. The disadvantages are that they are heavy, corrosive to reactor structural materials at higher temperatures, and lead-bismuth produces ^{210}Po as an activation product, which creates a significant radiation hazard. Another disadvantage is the very limited operational experience with lead/lead-bismuth reactors, which is based entirely on the experience of the former Soviet Union with several submarine propulsion units during the 1960s and 1970s.

4.3.4 Salt-cooled reactors

In the 1960s and 1970s, a completely different reactor type was developed in the US, resulting in the construction of two experimental reactors. This new reactor type used a molten fuel–salt mixture circulated through a graphite moderator block to achieve a very compact, high-power reactor system. Although originally intended for aircraft propulsion applications, which were abandoned in the early 1960s, the technology for MSRs continued to be developed for another 10 years as a candidate "breeder" reactor that could produce more fuel than it consumed.

Today, a number of countries are studying MSRs for potential commercial power or high-temperature process heat applications. Salt-cooled reactors offer opportunities to increase primary system temperatures above that available from LWRs while offering the benefits of liquid metal coolants such as good heat transfer characteristics and low primary system pressure. In addition, liquid salts are transparent, which improves inspection and maintenance operations relative to liquid metals, and the very high temperatures possible with salt-cooled reactors enable them to be used for industrial process heat applications.

The earlier MSRs were fueled with uranium or thorium fluorides dissolved in molten lithium and beryllium fluoride salts. Because the fuel was liquid, fission products could be removed from the fuel, and fresh fuel could be added while the reactor continued to operate. Systems for the chemical control of the molten salts were developed, but this technology has been stagnant for more than 30 years. A new variant of the MSR has emerged that uses pure molten fluoride salts, that is salt without fuel, to cool a graphite-moderated core containing standard graphite-coated particle fuel. This concept is referred to as a fluoride salt, high-temperature reactor (FHR) and takes advantage of the excellent thermodynamic characteristics of liquid salt to overcome many of the engineering challenges of the high-pressure, high-temperature GCRs. The use of solid fuel in the FHR avoids the design and regulatory challenges associated with the circulating fluid fuel in the MSR.

There has been a resurgence of interest in both the MSR and the FHR concepts. The MSR is one of the designated Generation IV concepts, and the FHR is a hybrid of the MSR and VHTR concepts. In collaboration with the US, China is currently designing a small FHR test reactor and expects to follow with a larger MSR test reactor.

4.4 Big versus small

To reiterate, SMRs are not a unique reactor technology but rather a design option for all technologies, and there are multiple examples of SMR designs using each of the technologies discussed in this chapter. The advantages and disadvantages of each technology, as discussed above and summarized in Table 4.1, apply to SMRs as well as their bigger siblings. The choice of size is influenced by the intended application

Table 4.1 **Key advantages and disadvantages of different nuclear reactor technologies**

Technology	Advantages	Disadvantages
Water reactors	• Large amount of design and operational experience • Common and benign coolant	• Low power conversion efficiency • High-pressure primary system
Gas reactors	• High-temperature coolant for process heat applications • High power conversion efficiency	• Limited amount of design and operational experience • High-pressure primary system
Liquid-metal reactors	• Improved fuel utilization and waste management • Low-pressure primary system	• Limited amount of design and operational experience • Difficult coolant handling issues
Salt reactors	• High-temperature coolant for process heat applications • High power conversion efficiency • Low-pressure primary system	• Very little design and operational experience • Coolant chemical compatibility issues • Licensing issues with liquid fuel

or by specific performance goals. For instance, a focus on markets with small electricity or process heat demand would dictate smaller units to better match the energy requirements. Achieving a specific total plant cost or a certain safety margin might also dictate a smaller unit size. There are many reasons to consider smaller-sized nuclear plants, especially safety, affordability, and flexibility. These considerations are discussed in depth in the next three chapters.

References

1. *A technology roadmap for generation IV nuclear systems*. GIF-002-00. December 2002.
2. Ingersoll DT, Poore III WP. *Reactor technology options for near-term deployment of GNEP grid-appropriate reactors*. Oak Ridge National Laboratory; 2007. ORNL/TM-2007/157.
3. Todreas N. Small modular reactors for producing nuclear energy: an introduction [Chapter 1]. In: *Handbook of small modular nuclear reactors*. Cambridge, UK: Woodhead Publishing; 2014.
4. *2015 Nuclear news reference special section*. Nuclear News, American Nuclear Society; March 2015.
5. *Reactor concepts manual*. US Nuclear Regulatory Commission. Available at: http://www.nrc.gov/reading-rm/basic-ref/teachers/03.pdf and /04.pdf.

Enhancing nuclear safety

5

In this chapter, I begin to explore the specific features of small modular reactors (SMRs) that distinguish them from large plants. In most cases, these features are intentionally employed in the designs for one or more of three primary goals: to further enhance the safety and robustness of a nuclear plant, to improve the affordability of nuclear power, or to expand its flexibility for broader applications. While these goals are valuable in their own right, they are also essential in meeting the needs and expectations of the broad range of customers that SMRs are intended to reach. Each of these goals will be addressed in separate chapters. Before discussing the first and most important feature, enhanced safety, I need to begin with some basic SMR terminology and clarifications. If after reading these next few chapters you still desire more technical detail on SMRs, I recommend that you consult the *Handbook of Small Modular Nuclear Reactors.*[1]

5.1 SMR terminology and basics

Most generally, the term "small modular reactor" is intended to mean a reactor that has a power capacity of nominally less than 300 MWe and is sufficiently small in physical size that it can be fabricated in a factory and transported to the plant location for installation and operation with multiple identical units. Regarding the choice of 300 MWe as the upper bound on output capacity, this comes from the long-standing terminology of the International Atomic Energy Agency (IAEA), which classifies "small" as less than 300 MWe. The Energy Policy Act of 2005 formalized this terminology in the US by allowing multimodule plants to be treated as a single entity with respect to indemnification and insurance liability so long as each module has a power level below 300 MWe.[2] With regard to the term "modular," the nuclear community has already seen confusion over this term by those who mistakenly identify it with "modular construction." Modular construction techniques are common in many industries and are beginning to be used in large nuclear reactor construction projects. The point of SMRs is to extend conventional modular construction to the nuclear reactor system itself so that it can be completely fabricated in a factory. By implication, an SMR plant should have multiple nuclear modules operating in unison, although some vendors have proposed SMR plants employing only a single nuclear module. In my opinion, this is similar to my grandson declaring that he has built a house using a single LEGO® block—possible, but not very creative.

As reviewed in the previous chapter, the term "technology" in the context of nuclear reactor types most often refers to the primary coolant used to remove heat from the reactor core. The choice of coolant typically influences the overall design and substantially impacts the choice of other materials in the reactor system, such as the fuel, fuel cladding,

Small Modular Reactors. http://dx.doi.org/10.1016/B978-0-08-100252-0.00005-7

core structures, and reactor vessel. The most common reactor coolants are water (light or heavy), gas (helium or carbon dioxide), and metal (sodium or lead alloys). SMRs are not a specific reactor technology since designs have been developed using each of these different coolants. Instead, SMRs are simply a design option within a selected reactor technology, that is, a choice by the designer to limit the output capacity of the reactor. This choice may or may not impact the arrangement of the plant components, but the fundamental technology of the design remains mostly a function of the choice of coolant.

Another distinction regarding SMRs is between those that are merely small versions of large plant designs and those that are what I like to call "deliberately small," that is, they are designs that cannot be scaled to large sizes but rather capitalize on their smallness to achieve specific performance goals. The integral primary system reactor configuration mentioned in earlier chapters is a good example of a deliberately small reactor feature and has been used in many early and current SMR designs. Most importantly, the integral configuration provides an opportunity for significantly improved plant safety, which is discussed in depth in this chapter, and plant simplification, which can enable competitive economics (discussed in the next chapter).

A confusing aspect of SMRs is that there are a vast number of potential applications for them proposed by both commercial and research organizations, especially for very small designs in the range of 1–20 MWe, sometimes referred to as "micro modular reactors." In addition to some of the exotic military uses of small nuclear reactors proposed by the US Air Force and Army in the early days, others have proposed SMRs for mobile emergency response power units, space travel, cruise ships, unattended "nuclear batteries," and just about any application that a creative mind can conjure. With this diverse range of applications comes an equally diverse range of designs tailored to address the unique requirements of the different applications. All of these applications of nuclear power are intriguing, and some may eventually happen. However, I will stay focused on more classical applications of electricity generation and heat delivery from stationary plants.

There is one class of SMRs worth mentioning that is different from the traditional land-based plants. This is the "transportable nuclear power plant." This classification refers to barge-mounted power units being developed by the Russian Federation to provide electrical power and heat to isolated communities along their Arctic Sea coast. These SMRs are derived from nuclear units that currently provide propulsion to Russia's fleet of icebreakers. They offer a near-term solution to power remote communities; in fact, the first barge with two small 35 MWe nuclear reactors is currently under construction in St. Petersburg. The only US experience with transportable nuclear power plants is the USS Sturgis, which was discussed briefly in Chapter 2 as part of the US Army's nuclear power program. The Sturgis contained a single 10 MWe nuclear reactor that provided electricity to the Panama Canal for several years.

5.2 Safety and the nuclear power industry

The safety of nuclear plants initially crept into the dialogue in the early 1970s, due in part to an industry-led study to evaluate the statistical probability of reactor accidents. Opponents were quick to seize this message and have been driving the dialogue in a negative direction ever since. The nuclear industry perpetuates this perception by being defensive about nuclear plant safety despite all evidence to the contrary. In reality, nuclear plants are incredibly safe. By any metric you choose, the nuclear industry has one of the best records among all industries for protection of the workforce and the general public.

Geologists Judith Wright and James Conca published a book in 2007 detailing the characteristics of various energy sources that are likely to be part of the US energy mix for the foreseeable future.[3] In their book, *The Geopolitics of Energy*, they address several of the common myths of nuclear power, including safety. The data in Table 5.1 are derived from their assessment of US fatalities in the period 2001–2006 and show the average annual number of deaths in the US resulting from the listed activities. "Iatrogenic causes" is a new term for me. These deaths result from unexpected and unexplained outcomes of medical procedures. It was not surprising that smoking, alcohol, and automobile accidents are high on the list—most people would expect that ranking. The fact that nuclear power is at the bottom of the list does not surprise me either, but I have to wonder how many people have a different perception of its risk.

As a second metric for safety, Wright and Conca also presented data from the Occupational Safety and Health Administration comparing nonfatal injuries in three different US industries: manufacturing, finance (including insurance and real estate), and nuclear power. This comparison is shown in Figure 5.1. Given that nuclear plants are large, complex facilities that require constant operational and maintenance activities on industrial-scale equipment, the very low injury rate for the nuclear industry is a very telling metric.

Conca presented data published in Forbes comparing global mortality rates from several different electricity generation technologies based on the number of direct deaths and epidemiological estimates from several health organizations.[4] The data, which are normalized by the total amount of electricity generated, show that coal yields 170,000 deaths/TWh globally (280,000 deaths/TWh in China) compared to 4000 deaths/TWh for natural gas, 150 deaths/TWh for wind, and 90 deaths/TWh for nuclear, including the accidents at Chernobyl and Fukushima. I present these numbers

Table 5.1 **Average annual US fatalities for the period 2001–2006[3]**

Activity	Number of deaths
Iatrogenic causes	190,000
Smoking	152,000
Alcohol	100,000
Automobile accidents	50,000
Firearms	31,000
Coal use	6000
Construction	1000
Hunting	800
Police work	160
Nuclear power use	0

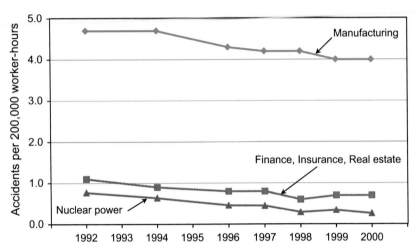

Figure 5.1 Occupational safety and health administration accident rates for three US industries (1992–2000).[3]

with some reservations because epidemiological studies must make numerous assumptions regarding the underlying relationship between a stimulus (e.g., radiation exposure or carbon inhalation) and consequential health effects. These assumptions may not be unanimously accepted and are sometimes hotly debated among the technical community. Although the precise numbers may be subject to debate, the relative numbers are likely indicative of the proportional risks associated with each of these energy sources.

Given the exceptional safety record of the nuclear industry, which has been consistently demonstrated over multiple decades of worldwide experience, I do not understand why many people, including industry advocates, persist with casting doubt on the safety of nuclear power. *It is time to move on.* I will concede, however, that there are two aspects of nuclear plant operations that are relatively unique and contribute substantially to the overall perception of risk. The first is radiation: that invisible, colorless, odorless threat to human health. I have studied and worked with radiation throughout my professional career, and I am quite comfortable with its risk. But I can appreciate the concerns from those who are not. The nuclear industry is committed to ensuring that workers and the public are not exposed to excessive amounts of radiation. Without digressing into the raging controversy regarding the correlation between low levels of radiation exposure and health effects, it is sufficient to acknowledge that the perception of risk is real, and we in the industry must respect that.

The second challenging aspect of nuclear plant safety is that reactor accidents typically occur over a protracted period—potentially weeks or even months. This is a result of the fundamental physics of the nuclear reactor core: it continues to generate heat long after the reactor is "turned off." This phenomenon was vividly demonstrated in Japan in 2011. The major earthquake and resulting tsunami that occurred on March 11 were largely over within hours, but four of the six nuclear reactor units that were damaged at the Fukushima Daiichi plant continued to capture the world's attention for many days, then weeks as emergency workers struggled to cool the damaged reactors.[5] The reactor fuel overheated, producing hydrogen from a chemical reaction between the fuel cladding and the coolant water. The hydrogen leaked into other parts of the plant and caused multiple explosions over the first several days of the accident. The overheated and exposed fuel released radiation into the cooling water, which leaked into the sea. Radioactive gases were also emitted directly to the atmosphere and contaminated a sizable area of land. Although the actual consequences from the reactor accident pale in comparison to the greater than 20,000 lives lost by the tsunami, the reporting frenzy sparked by the natural disaster eventually turned to the Fukushima accident that was still in progress. The resulting substantial press coverage of the protracted Fukushima accident helped to further exacerbate the general public's perception of nuclear power risks.

Alvin Weinberg, a pioneer in the development of nuclear power, compared nuclear power to a "Faustian bargain," offering much benefit but requiring much responsibility in its use. The entire nuclear industry takes this responsibility quite seriously and implements it through a broad network of domestic and international organizations that help to facilitate and monitor the safe operation of nuclear plants. The first and perhaps most important step in achieving plant safety is to design and engineer plants with the lowest possible risk. As I will discuss in the next sections, smaller-sized reactors, and especially deliberately small reactors, represent a major advancement in this direction.

5.3   Designing beyond safety

Plant safety is not at the discretion of the designer; it is an expectation set by the regulator. The designer is obliged to develop a design that meets these expectations in a convincing and substantiated manner. Safety standards established by the regulator are focused, as they should be, on reasonable assurance of adequate protection of the workforce and the public. The often cited concern for nuclear safety is not as much about its actual safety but rather a perception of its risk, which has both a technical basis and an element of emotion—sometimes irrational. For example, a person may be too afraid to board a commercial airliner but not hesitate to jump into the family car, a much riskier proposition.

The US Nuclear Regulatory Commission (NRC) sets an extremely high bar on its established safety standards and is frequently referred to as the "gold standard" for nuclear regulation worldwide. All reactor designs that have been certified by the NRC meet these exacting standards. The trend in new plant designs, large and small, is to meet the NRC safety standards in more assured ways, principally by making use of fundamental laws of physics whenever possible rather than relying on engineered systems that require operator action and a sustained source of electrical power to function. In doing so, the plant naturally becomes more resilient against unanticipated upsets.

Even though all existing plants have been properly licensed and therefore are considered "safe" according to NRC standards, it is desirable and appropriate to continue to enhance the level of safety in new plant designs. I recall deciding to install front seatbelts in my first car—a 1991 Ford Falcon—because it seemed like the safer thing to do. The last new car I bought had a full set of lap and shoulder belts in all passenger locations, 12 airbags, and a long list of other hidden safety features. Its safety rating was an important selling point.

Like cars, it makes sense to continually improve the safety of nuclear plants. But it may be instructive to think of it as designing beyond safety. This requires introducing new terminology that is a superset of "safety." I frequently use the term "robustness" to capture this quality. While "safety" relates to the protection of the workforce and the public, "robustness" relates to protecting the plant itself—ensuring that the financial investment and the generating capacity is not lost after a major upset. The accident at the Three Mile Island nuclear plant in Pennsylvania in 1979 resulted in no direct impact on the workers or the public (other than extreme anxiety); however, the damage to the reactor core caused a total loss of the unit and was a financial disaster for the plant's owner. Said differently, if a plant can survive a major upset intact, the protection of the workforce and the public will also be ensured. Realizing this distinction, many SMR designers seek to design for plant robustness, which is beyond safety.

Design choices impact its safety characteristics and also its robustness. Compare a delicate wine chalice and a beer stein: both are fundamentally used for serving an alcoholic beverage, but the stein is considerably more robust. Similarly, engineered systems can be designed with robustness in mind or not. Using the automotive analogy again, an off-road utility vehicle is designed with much more robustness than a finely tuned sports car. Heavy duty suspension, a reinforced frame, and perhaps even a roll cage help the off-roader to survive a wider range of use (and misuse). Many SMR designers understand this principle and design to the high-level goal of significantly enhancing the robustness of their design, even if they articulate it as enhanced safety. From my direct experience, Mario Carelli did this with his "safety by design" approach for the International Reactor Innovative and Secure (IRIS) design and Jose Reyes is doing this for the NuScale design. Others do as well, but since I have worked with both IRIS and NuScale developers and have closer familiarity with those designs, I better understand the motivations behind their design features. For the same reason, I will tend to refer more often to these two designs in my examples below for exemplary purposes only.

5.4 Designing for robustness

Commercial nuclear power plants are large, complex facilities that contain lots of pipes, pumps, valves, turbines, generators, and electrical equipment similar to coal and gas-fired power stations and other industrial facilities. Standard industrial safety practices are required throughout the plant. In addition, there is one hazard that is unique to nuclear plants: radiation. During normal operation, there are two primary sources of radiation: one direct and one indirect. Direct radiation (predominantly neutrons and gamma rays) results from the fission process within the reactor core. This hazard is very intense but is only significant close to the core and can be easily mitigated with shielding materials. The structural steels within the reactor system and even the water coolant itself provide effective shielding of the direct fission radiation. Indirect radiation results from material "activation," in which the fission neutrons cause a transmutation of normally stable material to unstable material and cause it to emit secondary radiation (mostly gamma rays). Although much less intense than the direct core radiation, the induced radiation can migrate via the circulating coolant to other locations in the plant, which creates a more distributed contamination hazard. In both cases, the hazard prevention mitigation strategies are well understood and are implemented as part of the plant design and operational procedures.

The real concern for nuclear power plants is when things go wrong—terribly wrong, such as at the Three Mile Island, Chernobyl, and Fukushima plants. During the opening plenary session at a nuclear conference, Nils Diaz, former NRC Chairman, spoke on the topic of nuclear plant safety and offered the following statement: "The only thing we have to do is cool the core—everything else is gravy."[6] He is referring to the reactor's decay heat, which I discussed briefly earlier in this chapter. If the persistent residual heat that continues to be generated long after the fission process is stopped is not effectively removed, then the reactor fuel and its protective metal cladding will heat up beyond acceptable limits. At this point, the cladding will fail, and the nuclear fuel will release an intense accumulation of radioactive material—solid and gaseous—into the reactor system and potentially into the environment. To paraphrase Diaz, reactor safety (and robustness) comes down to ensuring that the decay heat can be removed under all situations, anticipated or not.

Smaller-sized reactors offer more options to address this overarching issue of heat removal by the simple fact that there is less decay heat to remove. Different SMR vendors have chosen a variety of ways to capitalize on this advantage. Despite the number of design differences among vendors, many SMRs share a common set of design principles to enhance plant safety and robustness. The top-level principles include: (1) eliminate as many features as possible that have the potential to initiate a serious accident, (2) for those features that cannot be eliminated, reduce the probability that an accident will be initiated, and (3) design the system to substantially mitigate the consequences of remaining potential accidents. A comprehensive review of many design features for small- and medium-sized reactors spanning all reactor technologies is given in a 2009 report issued by the IAEA.[7] Some of the specific features that help to implement these principles and achieve a higher level of plant robustness are described below with an emphasis on water-cooled reactors, although not exclusively.

5.4.1   Passive safety systems

I have already used the term "passive" safety in previous chapters without much explanation. The term appears liberally in the following pages, so it may be useful to explain it more fully. The original generation of commercial nuclear power plants used "active" safety systems almost exclusively, that is, systems that require an operator action to initiate the system and some form of motive force such as electricity to perform the action. Beginning in the 1980s, there was considerable interest in designing safety systems that do not require operator initiation and electricity. The result was the development of passive systems that initiate and operate using only forces of nature such as heat conduction and gravity. Using my favorite analogy of the automotive industry, antilock brakes and power windows are active systems since the driver must press a pedal or button for them to operate. In contrast, the 5 mph bumper and collision-triggered airbags are passive systems. In a nuclear plant, backup feedwater pumps and motor-operated valves are examples of active systems. An example of a passive system is the use of natural convection of the coolant to remove heat from the reactor core during accident conditions or even for normal plant operation. Natural convection is fundamentally gravity at work—water becomes lighter and rises when heated and then becomes heavier and falls when cooled. Proper placement of the heat source and the heat sink produces a naturally circulating flow without pumps.

The formal definition of passive and active in the context of nuclear power plants is not precise and varies among authoritative sources. The IAEA provides a graded scale between fully passive and fully active systems, depending on the nature of the trigger used to actuate the system and the motive force used to operate it.[8] I will not add to the confusion by offering my own definitions, but suffice it to say that the desirability of passive systems over active systems is that they are generally more reliable—gravity is always present. This adds robustness to the plant since the designer does not have to deterministically plan for every conceivable upset condition. However, forces of nature can be relatively weak, and they are constant. This requires that the reactor design be tailored to the natural forces rather than vice versa. Also, passive systems tend to work better with small reactors where the heat generated in the reactor core is relatively small and manageable. This is not exclusively true, and there are clear examples of large reactors that make good use of passive systems, notably Westing-house's AP-1000 and General Electric's ESBWR. However, SMRs are able to extend this approach and to use almost exclusively passive systems to provide both safety and robustness to the design. You will see many examples of this below.

5.4.2 Arrangement of primary system components

The reactor designer can make a number of choices regarding the placement of key reactor system components that influence the overall robustness of the design. The most common and significant choice is to use an integral configuration in which all or most of the primary system components are contained within a single vessel. It is difficult to come up with a good automotive analogy; the best I can do is to suggest that the traditional loop-type configuration is similar to a motorcycle with one or more sidecars attached, while an integral-type configuration is like the family sedan with everyone tucked safely inside. The point is that the integral configuration eliminates many external vulnerabilities, in analogy to the potential detachment of a sidecar.

Several US and international SMR designs use an integral pressurized-water reactor (iPWR) design. This is a critical design simplification feature that is central to both the improved safety case and reduced plant cost. It is also the primary feature that keeps the reactor output deliberately small due to the limited volume within the primary vessel. Most importantly, the integral design eliminates the high-consequence accident scenario of a large pipe-break loss-of-cooling accident. It also greatly reduces the size of penetrations through the reactor pressure vessel, which therefore limits the rate at which coolant can be expelled from the vessel if one of those penetrations is breached. In a typical iPWR, the maximum size pipe penetrating the reactor vessel is 5–7 cm in diameter, which is needed for the feedwater inlet and steam outlet of the internal steam generator. This is in contrast to the 80–90 cm diameter pipes in a large loop-type PWR that connect the reactor vessel to the external steam generator vessels. A conceptual comparison of the two types of primary system configurations is given in Figure 5.2.

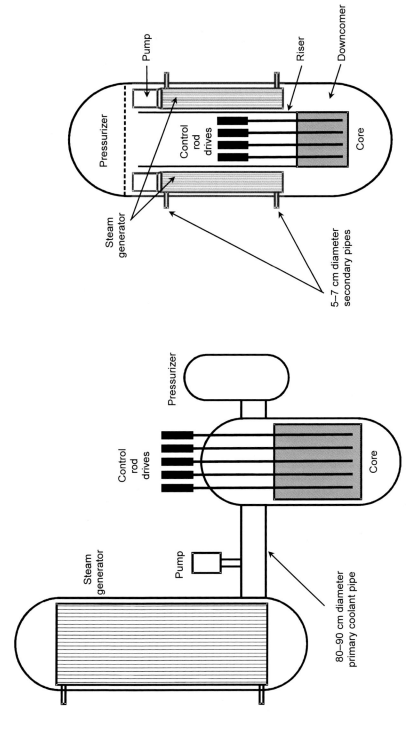

Figure 5.2 Comparison of loop-type (left) and integral-type (right) PWRs, showing the elimination of large primary coolant pipes.

The first (and only) commercial iPWR was aboard the NS Otto Hahn merchant ship, which was commissioned in 1968 and powered by a 35 MW integral reactor.[9] In the late 1980s, the 320 MWe Safe Integral Reactor (SIR) design was developed by a consortium including the United Kingdom Atomic Energy Agency, Combustion Engineering, Stone and Webster, and Rolls Royce.[10] The SIR design was developed specifically to respond to the safety challenges encountered by the early large loop PWRs and represents a precursor to many of the SMR designs that have emerged in the past few years. Nonwater-cooled SMRs may also use the integral configuration. In the case of metal and salt-cooled reactors, this configuration is more commonly referred to as a "pool" configuration, but it is functionally the same as an integral configuration.

Packing all of the primary system components into a single vessel has a number of safety-related advantages:

- All large coolant pipes are eliminated. Only small (5–7 cm diameter) feedwater and steam outlet pipes penetrate the primary vessel wall compared to large (80–90 cm diameter) pipes for loop-type PWRs. The smaller-sized penetrations significantly reduce the rate at which primary coolant water can escape from the vessel after a pipe break, thus delaying the potential consequences from a loss of coolant accident.
- Typically the heat exchangers are placed above the core elevation, creating a relatively tall system that facilitates more effective natural circulation of the primary coolant in the case of a coolant pump failure. Some designs have sufficient natural circulation flow rates to eliminate the primary coolant pumps entirely and rely only on natural circulation for normal operation. This approach completely eliminates accident scenarios associated with pump failure.
- All of the primary coolant is contained within a single vessel, which must be relatively large (tall) to accommodate all of the primary system components. This results in the amount of water inside the vessel being much larger per unit power than for an external loop-type PWR and increases the overall heat capacity and thermal inertia of the system. This in turn yields a slower system response to core temperature transients and allows the plant control system and operators more time to respond to the transient. For instance, the primary vessel water inventory in a NuScale module is roughly four times larger per unit power than for a traditional loop PWR.
- The presence of the steam generators within the primary vessel provides an internal, readily accessible heat sink for decay heat removal and facilitates the implementation of passive heat removal options in the secondary system.
- The extended riser area above the core provides the possibility for internal placement of the control rod drive mechanisms (CRDM), thus eliminating another potentially serious accident scenario: the rod ejection accident in which a control rod is inadvertently propelled upward out of the reactor core. With internal CRDMs, the control rod movement is constrained by the upper vessel internal components. Internal CRDMs also reduce the number of penetrations in the reactor vessel head, which reduces the likelihood of the near-accident situation that occurred at the Davis-Besse plant in 2008. In this case, boric acid in the coolant water leaked through a control rod penetration seal and corroded the vessel base material.[11] Only a thin stainless-steel inner liner prevented a major breech in the vessel pressure boundary. Some iPWR designs take advantage of this feature while others use external CRDMs.
- In some iPWR designs, a large downcomer region exists between the reactor core and the primary vessel due to the relative placement of the steam generators radially outward from the reactor core. The additional water in this region provides effective shielding of the direct radiation and results in a lower radiation exposure of the reactor vessel. This reduces the activation level of the vessel material and reduces another major safety concern: pressurized thermal shock, which results from radiation-induced embrittlement of the reactor vessel.

Table 5.2 **Comparison of reactor vessel dimensions for a traditional loop-type PWR and several current iPWR SMRs**

Parameter	Loop-type PWR	Integral PWR designs		
		NuScale	mPower	W-SMR
Height (m)	13.4	17.4	25.3	24.7
Diameter (m)	4.6	2.9	3.9	3.5
Aspect ratio	2.9	6.0	6.5	7.1

The major downside of the integral design is that it constrains the reactor to relatively small power levels, that is, it forces the reactor to be "deliberately small." Beyond some practical limit, probably in the 300–400 MWe range, the size of the vessel would become prohibitively large to manufacture and transport. To compensate for its inability to scale dimensionally, iPWR designers depend instead on scaling by replication, hence the interest in small *modular* reactors. An example of this is the NuScale design,[12] which uses twelve 50 MWe iPWR modules to comprise its reference plant design, and the Babcock & Wilcox mPower design,[13] which uses two or four 180 MWe modules to comprise its SMR plant.

Designing the primary system components with a high aspect ratio enhances safety and robustness by facilitating natural circulation of the primary coolant. In the case of iPWRs, accommodating all of the primary system components in a single vessel, while also constraining the vessel diameter to truck or rail transport limits, results in the iPWR reactor vessel being proportionally taller than for a loop-type PWR. For example, the vessel height-to-diameter ratio for a typical large PWR is roughly 3.0 and for a large boiling water reactor (BWR) is about 2.0. In contrast, the NuScale and mPower designs have aspect ratios in the range of 6.0–6.5. This increase in the aspect ratio enhances the formation of gravity-driven natural circulation of the coolant, which enhances heat removal from the core and provides an effective means of transferring the heat to what is called the "ultimate heat sink" even if power to drive the coolant circulation pumps is lost. In some iPWR designs such as NuScale, the natural circulation driving force is designed to be sufficiently strong to be used as a core cooling mechanism for full power operation, thus eliminating the need for pumps entirely. The Holtec SMR-160 design also uses natural circulation for normal operation even though it is not strictly an iPWR configuration. It does so by placing the steam generator in a very tall vessel that is above and tightly coupled to the separate reactor vessel. Table 5.2 summarizes the aspect ratio parameters for three US iPWR designs compared to a traditional loop-type PWR.

5.4.3 Decay heat removal

As discussed earlier in the chapter, the protracted generation of decay heat in a reactor core is a distinguishing feature of nuclear power plants and the driving force behind most plant safety and robustness considerations. Reactor decay heat must be removed from the reactor core for an extended period of time after the reactor is shut down

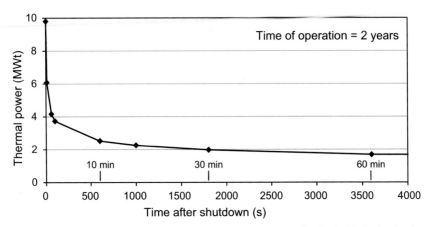

Figure 5.3 Decay heat as a function of time after the shutdown of a single NuScale-sized module that has been operating continually for 2 years.

to avoid fuel damage. If the fuel is not damaged, the intensely radioactive materials within the fuel will not be released. The amount of decay heat power is roughly proportional to the reactor's full power capacity; therefore, a 150 MWe reactor will have one-tenth the amount of the decay heat power of a 1500 MWe reactor. It is important to note that even though a small reactor module will have lesser decay heat power than a large reactor unit, the decay heat can still be significant, and fuel damage will occur unless the heat is adequately dissipated through one or more removal paths. Having less of it, however, opens up a lot of design options for effectively removing it—a key principle in SMR designs.

The bad news is that the decay heat lasts for a long time. The good news is that it drops to low levels very quickly. In Figure 5.3, I have estimated the approximate decay heat in a single NuScale-sized module for the first hour after being shut down subsequent to 2 years of full power operation. The curve was generated by assuming a full power of 160 MWt and using a classic textbook formula for the decay heat power.[14] Within the first second, the thermal power drops to roughly 6 MWt, equivalent to 6% of the full power level. At 1 hour, the thermal power is less than 2 MWt and continues to drop slowly with time. Using the same formula, the power is less than 400 kWt after 30 days. Kent Welter, NuScale Safety Analysis Manager, likes to compare this amount of heat to "a couple hundred hair dryers."

Because of the lower full power rating of the SMR core, the decay heat will be proportionally smaller to a level at which it is easier to remove the heat with natural means. As discussed above, the tall vessel configuration of many SMRs enhances natural circulation of the coolant through the core, which transfers the heat to the internal heat exchangers or the vessel surface. Also, the smaller core volume enables more effective conduction of the decay power to the reactor vessel. This is a result of the smaller core radius, which results in a shorter conduction path from the core centerline (the hottest part of the core) to the reactor vessel. Many SMRs have cores of fewer than 70 standard fuel assemblies compared to as many as 250 assemblies in a large

Table 5.3 **Comparison of reactor vessel surface area and power rating for a loop-type PWR and several current iPWR SMRs**

Parameter	Loop-type PWR	Integral PWR designs		
		NuScale	mPower	W-SMR
Power (MWe)	1200	50	180	225
Surface area/ volume (m^{-1})	1.02	1.49	1.10	1.22
Surface area/ power[a]	1.0	18.4	9.1	7.2

[a]Relative to traditional PWR.

PWR and greater than 800 assemblies in a large BWR. Even with these improved heat convection and conduction paths to the vessel surface, additional external cooling of the vessel is needed to transfer the heat to the ultimate heat sink, typically the ground or atmosphere.

Fortunately, the removal of heat from the external surface area of the vessel is more effective for smaller systems—a benefit derived from basic geometry. This advantage results from the fact that core power, and hence decay power, is proportional to the volume of the core, which is a function of the cube of the effective core radius. On the other hand, heat removal from the exterior surface of the vessel is proportional to the vessel surface area, which is a function of the square of the vessel radius. As the system size shrinks, the core volume decreases faster than the surface area of the vessel, or stated more simply, the surface-to-volume ratio increases. As the surface-to-volume ratio increases, the relative efficiency of external heat removal via the vessel surface also increases. As a result, most small reactor designs are able to easily accommodate removal of the decay heat using fully passive, natural convection water or air circulation to cool the outer surface of the vessel. Table 5.3 provides a comparison of vessel surface area and power characteristics for a standard loop-type PWR and the same three US iPWR designs given in the previous table.

5.4.4 Other design features and options

There are a number of other SMR design features and options that can enhance the plant's safety or robustness and are a result of the reactor's smallness. Most importantly, the radionuclide inventory in the reactor core, which represents the dominant radiation hazard in a nuclear plant, is roughly proportional to the reactor's power level. Therefore, a 150 MWe reactor will have one-tenth the amount of radionuclides in the core compared to a 1500 MWe reactor. The amount of radiation hazard assumed to be released in an accident, referred to as the "source term," is a combination of the radionuclide inventory and potential release paths. In addition to the intrinsically smaller radionuclide inventory of an SMR, some SMR designs add additional barriers to reduce and delay fission-product release to achieve a dramatically smaller accident source term. These might include immersing the containment vessel in water and surrounding the multiple containments within a multimodule plant with an additional protective building. Multimodule plants, that is, plants that are comprised of several SMR units, will have a proportionately larger radionuclide inventory, depending on the number of modules. However, it is unlikely that the collective source term scales linearly with the number of modules. This would only be the case if all modules underwent the exact same accident scenario and experienced the exact same failures of the defense-in-depth systems. In reality, the likelihood of this occurrence is very low. A real-world example of this is at the multiunit Fukushima plant in Japan. Four of the six units were subjected to the same earthquake and tsunami conditions, but there were considerable variations in the severity and timing of system failures in the various units.

The smaller plant footprint of some SMR designs makes it more economically viable to construct the primary reactor system fully below ground level, which significantly hardens it against external impacts such as aircraft or natural disasters. As an example, the W-SMR design has a containment vessel volume that is more than 23 times smaller than the Westinghouse AP-1000 containment, which allows for a significantly smaller footprint of the plant's safety systems.[15] In addition to hardening the safety systems to external impacts, below-grade construction helps reduce the number of paths for fission-product release in the event of a severe accident.

Core geometry can also contribute to a more robust design. In some cases, merely reducing the size of the core (and power) may not be sufficient to provide full passive removal of the decay heat. Such is the case with the high-temperature gas-cooled reactor (HTGR), in which helium is used as the primary coolant rather than water. The power level of General Atomics' original HTGR design was 2240 MWth but was later downsized to 600 MWth to improve its safety case.[16] Even at this power level, however, it was difficult to ensure adequate decay heat removal using only natural laws of physics, owing largely to the low heat capacity of the helium coolant. For example, during a loss-of-forced-flow accident, natural circulation of the helium coolant is able to remove some but not all of the decay heat. The rest must be conducted to the reactor vessel. The relatively large diameter of the core creates a very lengthy radial conduction path from the center of the core to the vessel surface. This results in unacceptably high temperatures at the core centerline. Instead of reducing the outer diameter of the core, General Atomics adopted an annular core design for its Modular High-Temperature Reactor (MHR),[17] in which the central portion of the core is replaced by a nonfueled graphite moderator, as shown in Figure 5.4. The annular core design significantly shortens the heat conduction path between the core hot-spot and the reactor vessel, and the additional graphite moderator in the central region increases the thermal inertia of the system, yielding a more robust response to loss-of-forced-flow accidents. With this core configuration, it is possible to remove the decay heat using passive cooling on the external surface of the reactor vessel.

The traditional approach within steam generators is to circulate the primary coolant through the inside of the tubes of the steam generators and boil the secondary water on the outside of the tubes (called the shell side). Since the primary water is at a higher pressure than the secondary water, the tube-side primary flow places the tubes

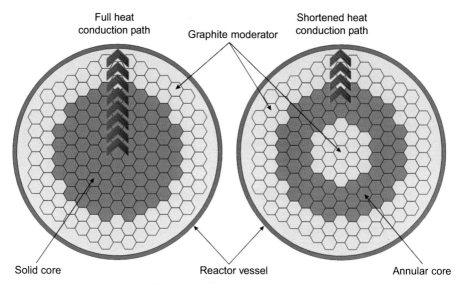

Figure 5.4 Comparison of solid core HTGR (left) and annular core MHR (right) with the improved use of passive decay heat removal.

in tensile stress and can result in tube rupture, which introduces potentially radioactive primary water into the secondary side of the plant. Some SMR designs such as IRIS and NuScale reverse this arrangement. By reversing the primary and secondary flows in the steam generators, the tubes now operate in compression, thus reducing the likelihood of a tube rupture, although introducing the possibility of a tube collapse. This design choice is expected to reduce the anticipated failure rate of the steam generators and also reduce radiological exposures during steam generator inspection, repair, and replacement activities.

A few SMR designs, most notably the NuScale design, operate with a vacuum region between the reactor vessel and the containment vessel. During normal operation, the vacuum region acts like a thermos bottle and obviates the need for vessel insulation, which has become a challenging concern in traditional plants. The concern is that insulation degrades with time and begins to disintegrate, causing insulation fibers to collect at the bottom of the containment building. The fibers can then clog sump filters and prevent water circulation during an accident when the containment water is needed to remove decay heat. Using a vacuum inside the containment vessel, which is only tractable for small-volume containments, completely eliminates this sump blockage concern.

Finally, some water-cooled SMR designs choose to remove soluble boron from the primary coolant as an added safety feature. Boron is added to the primary coolant in traditional large plants to compensate for the depletion of the nuclear fuel and the accumulation of neutron-absorbing fission products over the lifetime of the fuel. Eliminating the soluble boron has two important advantages: (1) it provides a strongly favorable reactor control feedback mechanism if the coolant water is voided for any reason such as unexpected boiling or a flow blockage, and (2) it avoids issues associated with boric acid corrosion similar to what occurred at the Davis-Besse plant. However, if soluble boron is not used, some other form of reactivity control must be provided such as solid burnable poisons or additional control rods. The safety implications of the alternative approaches must be carefully evaluated.

The cumulative impact of the added resilience provided by these collective design choices is that the individual reactor modules and the entire SMR plant should be able to survive a broader range of extreme conditions. In turn, this will help to ensure the safety of the workforce and the general public, and it will also help protect the plant owner's investment and continued availability of the power-generating asset.

5.5   Resilience to Fukushima-type events

The nuclear industry continually incorporates operational experience to improve the safety and efficiency of current and new plants. This is especially true regarding lessons learned from major accidents or near-misses. The destruction of four commercial nuclear units in 2011 at Japan's Fukushima Daiichi plant has already had a significant impact on the safety and emergency preparedness of existing plants worldwide. As an example, within weeks after the Fukushima accident, the European Council requested that a detailed safety and risk assessment be conducted for all nuclear plants in the European Union. The Western European Nuclear Regulators Association quickly assembled specifications for what became known as the nuclear "stress test," which was subsequently extended to reactors in several countries beyond the European Union.[18] In the US, the Nuclear Energy Institute coordinated an industry effort to develop an approach for adding to existing plants diverse and flexible strategies— referred to as FLEX—to mitigate the consequences of an extreme event. The FLEX program includes the addition of portable emergency response equipment and procedures for storing, protecting, and dispatching the equipment during an emergency.[19]

The American Nuclear Society issued an excellent summary of the Fukushima accident and its many implications in a special committee report published approximately 1 year after the accident.[5] The accident brought into focus four major considerations that are especially pertinent to SMR development and deployment. Because the SMRs are still in the design phase, they are well positioned to maximize the benefits of the lessons learned from the Fukushima accident. The key considerations that have special relevance to SMRs include the following:

1. Low-probability events can still happen.
2. The reliance on electrical power to prevent fuel damage is a substantial vulnerability.
3. Designing for a grace period of 24h, or even 72h, may not be enough.
4. Plants with multiple reactor units need to consider the potential impacts of accident progression from one unit to the others.

I will address each of these important considerations in order. First, regarding low-probability events, the regulator requires that the reactor designer convincingly demonstrates that the plant can survive a rigorous set of upset conditions. This collective set of postulated upset conditions is referred to as the plant's "design basis." The Fukushima plant experienced conditions that exceeded its design basis, specifically the 15-meter-high tsunami wave, resulting in a "beyond design basis" accident situation. It is not practical to design a plant to survive all conceivable events, no matter how unlikely—an extreme example is a direct meteorite hit. The challenge is designing enough robustness into a plant that it has a good chance of surviving even unexpected severe accidents while retaining the plant's economic viability. Traditionally, this additional level of robustness has been achieved by employing multiple, redundant layers of backup systems, referred to as "defense-in-depth." Most SMRs retain the defense-in-depth strategy while providing additional robustness by eliminating design vulnerabilities and relying on natural forces to operate the backup systems.

It is difficult to define metrics that adequately measure the robustness of a plant in very low-probability situations. A common metric is the "core damage frequency" (CDF), which is the statistical likelihood that an accident sequence will result in significant core damage. The NRC requires that a new reactor design have a CDF of less than 0.0001, that is, core damage would be expected to occur less than once in 10,000 years. The newest large plant designs appear to have CDFs that are one to two orders of magnitude lower (less likely). Several of the SMR designs under development quote values that are another one or two orders of magnitude lower, that is, one probable core damage event in nearly one billion years. I call that robustness. If the new SMRs can deliver on this claim, they will be substantially more resilient to severe conditions such as those experienced by the Fukushima plant.

The second consideration emphasized by Fukushima is associated with the importance of having assured backup electricity in the case of a reactor accident. This vulnerability is well known, and designers provide redundant layers of backup power options, including emergency diesel generators and safety-grade batteries—lots of them. The power is needed to run pumps that circulate the cooling water and to operate valves that direct the various sources of cooling water to the required locations. A more elegant solution to this plant vulnerability is to eliminate the need for backup power, at least for maintaining the reactor core cooling function. Because of the smaller amount of decay heat within an SMR, it is practical to design the reactor system to not require backup power for an extended and potentially unlimited time. At least two SMR designs claim to accomplish this. In the NuScale design, with which I am most familiar, this is accomplished by eliminating pumps entirely from the primary reactor system and immersing the module directly into a large pool of water to provide assured flow of heat from the reactor core to the ultimate heat sink.

The third consideration, which is somewhat related to the previous issue, is the broader concept of "grace period." The grace period is the length of time after an accident during which no external power, makeup water, or operator action is required to maintain the plant in a safe condition. Traditionally, 24h was used as the acceptable grace period, although newer plants are expected to provide a grace period of 72h. The accident at Fukushima demonstrated that more than 72h may be needed in severe accident situations. The devastating tsunami made access to the plant by emergency workers very difficult. The few workers that were available were heavily burdened by the many actions needed to deal with the accident progression in four large reactor units. In recognition of this challenge, and due to the smaller heat loads and radionuclide inventories in SMRs, several designs appear to have achieved severe accident grace periods of several weeks. At least one design (NuScale) offers an unlimited grace period, owing largely to its small module size of only 50MWe and the fail-safe nature of its simplified emergency core cooling system.

Whereas the first three considerations tend to favor SMRs due to their added robustness, the fourth consideration, that is, the implications of cross-unit interactions in multimodule plants, emphasizes a potential challenge for some SMRs. This issue manifested itself in two ways at Fukushima: (1) the limited workforce available at the plant was overburdened trying to manage the four reactor units that were experiencing extreme upset conditions, and (2) hydrogen that was produced from overheating of the core in one unit seeped into an adjacent unit, causing an explosion there. In the immediate aftermath of the accident, I had a number of colleagues suggest that the experience at Fukushima might preclude further consideration of multimodule SMRs. Fortunately, a more thoughtful response has evolved, and new methodologies are being developed to analyze multimodule effects. I would assert that a fundamental failing in the Fukushima experience is that the Fukushima Daiichi plant was never treated as a six-module 4700 MWe multimodule plant. Instead, it was treated as six individual plants with independent risks. Multimodule SMR vendors must fully address the issue of potential module-to-module codependencies and interactions throughout the design phase. My observation is that the vendors were already sensitive to this issue prior to Fukushima and that the experience in Japan served only as a vivid reminder of the consequences of not doing so.

5.6   Closing remarks on safety

There is so much more that could be said about the safety of nuclear power plants and the additional level of robustness that SMRs have to offer. Personally, I feel that enhanced robustness is the most significant benefit that SMRs can bring to the nuclear industry. Moving the dialogue from "safety" to "robustness" will be an important step in developing the new methods and metrics that will allow the industry to confidently advance the resilience of new plant designs against a broader range of unexpected threats—natural and man-made.

Most of the many earlier studies have similarly concluded that SMRs can provide exceptional safety and robustness. The consistent concern in these studies, however, has been that the enhanced safety comes at a price, and that price is economic competitiveness and viability. I disagree with this popular conclusion, and in the next chapter, I will present my case for why improved economics is a significant benefit that SMRs bring to the nuclear industry.

References

1. Carelli MD, Ingersoll DT. *Handbook of small modular nuclear reactors*. Cambridge, UK: Woodhead Publishing; 2014.
2. *Energy policy act of 2005*. United States Government; August 8, 2005. Public Law 109–58.
3. Wright J, Conca J. *The GeoPolitics of energy: achieving a just and sustainable energy distribution by 2040*. North Charleston, SC: BookSurge Publishing; 2007.

4. Conca J. *How deadly is your kilowatt?*. Forbes; June 6, 2012. Available at: www.forbes. com/jamesconca/2012/06/10/energys-deathprint-a-price-always-paid/.
5. *Fukushima Daiichi: ANS committee report*. American Nuclear Society; March 2012.
6. Diaz N. *Introductory remarks at the international congress on advanced power plants 2014*. April 2014. Charlotte, NC.
7. *Design features to achieve defense in depth in small and medium reactors*. International Atomic Energy Agency; 2009. NP-T-2.2.
8. *Safety related terms for advanced nuclear plants*. International Atomic Energy Agency; 1991. TECDOC-626.
9. Nuclear ship 'Otto hahn'. *Atomwirk Atomtech* 1968;**13**:294–330.
10. Matzie R, et al. Design of the safe integral reactor. *Nucl Eng Des* 1992;**136**:73–83.
11. *Davis-Besse reactor pressure vessel head degradation: overview, lessons learned, and NRC actions based on lessons learned*. U.S. Nuclear Regulatory Commission; 2008. NUREG/BR-0353, Rev. 1.
12. Reyes Jr JR. NuScale plant safety in response to extreme events. *Nucl Technol* May 2012;**178**(2):153–64.
13. Halfinger JA, Haggerty MD. The B&W mPower scalable, practical nuclear reactor design. *Nucl Technol* May 2012;**178**(2):164–9.
14. Glasstone S, Sesonske A. *Nuclear reactor engineering*. New York: Van Nostrand Reinhold; 1967.
15. Memmott JJ, Harkness AW, Wyk JV. Westinghouse small modular reactor nuclear steam supply system design. In: *Proceedings of the international conference on advanced power plants (ICAPP), Chicago, IL*. June 24–28, 2012.
16. LaBar MP. The gas turbine–modular helium reactor: a promising option for near term deployment. In: *Proceedings of the international congress on advanced nuclear power plants, Embedded Topical American Nuclear Society 2002 Annual Meeting, Hollywood, FL*. June 9–13, 2002. GA-A23952, 2002.
17. Shenoy A, et al. Steam-cycle modular helium reactor. *Nucl Technol* May 2012;**178**(2):170–85.
18. *Peer review report: stress tests performed on european nuclear power plants*. European Nuclear Safety Regulators Group; April 25, 2012.
19. *Diverse and flexible coping strategies (FLEX) implementation guide*. Nuclear Energy Institute; April 2012. NEI 12–06, Draft Rev. 0.

Improving nuclear affordability

Traditionally, economics has been viewed as being second only to safety in establishing the viability of nuclear power as a major contributor to the global energy portfolio. In light of the evidence presented in the previous chapter regarding the extraordinary safety record of the nuclear power industry and the potential to further enhance safety with small modular nuclear reactors (SMRs), I would conclude that economics has now replaced safety as the greatest challenge and uncertainty for expanding the use of nuclear power. The concern regarding uncertain economic viability is further exacerbated for SMRs due to a lack of direct experience with their construction and operation and a prevailing mindset that "bigger is better" (or at least cheaper). Perhaps this is why there have been several studies in recent years generating a plethora of papers and reports dedicated to proving or disproving the economic viability of SMRs. The good news is that there was no shortage of resources for crafting this chapter; the bad news is that the results and conclusions from the many papers and reports are quite diverse, sometimes contradictory, and frequently lack a clear understanding of SMR fundamentals.

Preparing this chapter presented a special challenge for me for two primary reasons: (1) I have no formal training in economics and limited experience with the subject, and (2) I admit to having a generally negative bias toward the value of economic studies, especially those that attempt to predict future costs and market potential. Despite this bias, I have spent considerable time researching the literature and even participating in a few economic studies in order to better understand this complex topic. My motivations were partly because of the importance of economics to the ultimate success of SMRs and partly because I frequently get quizzed on this subject when lecturing on SMR technology. While my understanding of economics has improved, my generally negative bias has not.

Several economic studies focus on predicting what the price of SMRs will be in the distant future and consequently their marketability. The problem with the projections, however, is that the predictive models are necessarily dominated by complex assumptions and large-group approximations. In their defense, the economists' task is virtually impossible—more difficult than trying to predict what the weather will be 5 years from today. This is because weather is (relatively) insensitive to human activities, while economics is quite sensitive to this highly unpredictable factor. A good demonstration of this vulnerability surfaced at a symposium that I attended in 2012. The symposium, which explored similarities and differences between natural gas and SMRs for electricity generation, included a well-recognized economist from the natural gas community. During his presentation, he openly admitted that the rapid deployment of "fracking" methods for enhanced gas recovery resulted in a dramatic drop in natural gas prices that was completely unanticipated only a few years ago and

Small Modular Reactors. http://dx.doi.org/10.1016/B978-0-08-100252-0.00006-9

negated all economic projections. I respected his honesty until he proceeded to confidently predict the price of natural gas in 20 years based on his economic model—the same model that failed miserably to predict current prices.

For these reasons, I offer no predictions of the anticipated costs of SMRs in this chapter either in terms of their capital and operating costs or the resulting cost of the electricity they produce. Instead, I will provide a more comparative analysis of SMR economics and focus on the reasons why I believe that SMRs will ultimately prove to be both affordable and competitive with the alternatives. Although the true economics of SMRs are yet to be demonstrated, I will offer the basis of my assertion that economics are a major strength of SMRs and not a handicap, as many assume.

6.1   The business of nuclear power

In the opening plenary session of the American Nuclear Society's Utility Working Conference in 2011, John Rowe, departing chief executive officer of Exelon Corporation, offered a frank challenge to the nuclear community:

> *The country needs nuclear power if it is going to tackle the problem of climate change, clean up our generation stack, maintain reliability and improve overall energy security. But we must keep our hopes for new generation harnessed to facts. Nuclear needs to be looked at in the Age of Reason and not the Age of Faith. It is a business and not a religion.*[1]

His comments were taken out of context by some to conclude that he had turned away from nuclear power as a viable option. But his remarks were really intended to bring reality to those who get so enamored with their favorite technology that they aggressively promote it without regard to potential shortcomings and frequently discredit the alternatives. Rowe explicitly cited politicians as falling into this trap, but I have seen many scientists and engineers suffer from the same myopic view.

Alvin Weinberg, a pioneer in the development of nuclear power and a staunch promoter of new technologies, also recognized the importance of economics. In his book *The Second Nuclear Era*, he wrote: "Unless a forgiving reactor is affordable, no one will buy it."[2] He was speaking specifically about the new smaller-sized, highly robust reactor designs that were emerging during the late 1970s and early 1980s. Many subsequent studies of smaller-sized reactor designs by the International Atomic Energy Agency (IAEA) and Nuclear Energy Agency (NEA) also identified potentially unfavorable, or at least uncertain, economics as the key concern for smaller-sized reactors.

With the rapid reemergence of interest in SMRs in the past 5 years, both the IAEA and the NEA have taken a fresh and more refined look at the anticipated economics of SMRs. In addition, several research organizations in the US and internationally have conducted studies related to SMR economics with varied emphasis on market potential, cost projections, economic competitiveness, and economic impacts. An excellent overview of SMR economic considerations is given in a paper entitled "Small modular reactors: A comprehensive overview of their economics and strategic aspects" by

Locatelli and colleagues.[3] Boarin and colleagues also provide a good foundation of SMR economic principles in Chapter 10 of the *Handbook of Small Modular Nuclear Reactors*.[4]

6.2   Rethinking economic metrics

A phrase that stuck with me from some of my earlier management classes is "you are what you measure." This is a clever attempt at drawing attention to the fact that the performance of an employee or an organization is influenced by the choice of metrics that are established to measure and reward progress. The challenge is choosing metrics that are consistent with both the goals of the organization and the skills of the employee. Inappropriate metrics can result in either the appearance of failure despite a high-quality effort or good progress in the wrong direction. A classic example comes from my experience at a national laboratory, whose primary role is to conduct impactful research. A prevailing metric used to evaluate personal and organizational success is the number of peer-reviewed publications and resulting citations, similar to academic researchers. While this works well for many divisions of the laboratory, those groups that work on classified programs tend to suffer at performance review time. Eventually, researchers avoided working on those programs, which challenged the laboratory to sustain this vital part of its mission. Eventually, a different and more appropriate metric was developed to evaluate researchers in those programs: the rate of repeat funding from sponsoring agencies.

A similar phenomenon exists with the emergence of innovative products. Earlier, I discussed how critics of Apple's iPad prematurely declared it to be a failure because critics were judging it against the same metrics used to evaluate smart phones in terms of portability or desktop computers in terms of computing power. By those metrics, it should have failed miserably. The reality was that it represented a new class of personal devices, which required the establishment of new metrics. The same situation is occurring with SMRs, especially with respect to their economics.

From the nuclear power industry's infancy, the dominant metric for evaluating the economic viability of nuclear power plants has been "levelized cost of electricity" (LCOE). Roughly speaking, this is the total life cycle cost of the plant normalized by the total energy production of the plant and is usually expressed as cents per kilowatt-hour (or dollars per megawatt-hour). This metric continues to be appropriate for large-capacity plants, which are typically located on extensive, interconnected grids with a robust wholesale electricity market. In this environment, the unit cost of electricity from each plant must be competitive with electricity generated by other plants—nuclear or otherwise—in the same wholesale market. While the LCOE metric makes sense for large plants, it is not obvious that it is equally important for SMRs, especially for those that are located on smaller regional grids or used for off-grid applications. Even for large-grid applications, it is necessary to expand the set of economic metrics in order to capture the new features that SMRs bring to the market such as incremental capacity growth and flexibility of operations.

A few studies have acknowledged this issue, although only in passing. For example, buried in Appendix C of a 2011 report by Rosner and Goldberg, *Small Modular*

Reactors—Key to Future Nuclear Power Generation in the US, is the following statement: "Understanding the economics of SMRs…may require the application of additional economic metrics for assessing economic feasibility and financial viability."[5] They go on to offer a few candidates, including capital-at-risk, which is the total capital cost that has not been repaid at any point in the project, and annual net cash flow, which reflects the time to initial cash flow and the periodic debt/equity profile. Another 2011 report by the NEA, *Current Status, Technical Feasibility and Economics of Small Nuclear Reactors*, hinted at additional metrics in a brief statement:

> *The attributes of small modular reactors, such as small upfront capital investments, short on-site construction time (with the accordingly reduced cost of financing), and flexibility in plant configuration and applications are attractive for private investors.[6]*

Unfortunately, the bulk of the economic assessment within the report focuses on comparing the LCOE of SMRs versus large reactors (LRs) to determine their economic viability in established markets—the type of analysis that is comfortable and familiar to traditional economists.

My surmise of the key financial parameters that are important to a fair assessment of the economic viability of SMRs is that the list should include at a minimum the following metrics:

- *Upfront capital cost.* This metric is basically the purchase price of the nuclear plant. It is central to the affordability of SMRs and is perhaps the leading discriminator between customers who can consider building a nuclear plant and customers for whom this is not an option.
- *Interest during construction.* This metric is a subset of the total project cost and represents the cost of borrowed money. It includes several considerations such as availability of money, credit rating of the owner, length of construction, total debt, and overall project risk.
- *Maximum cash outlay.* This metric, similar to capital-at-risk, is the largest amount of debt that has to be financed at any time during the project and influences the interest rate of borrowed money.
- *Time to first revenue.* This metric helps to capture investor risk concerns, especially return-on-investment and financing costs.
- *Sensitivity to market changes.* This metric also captures investor risk concerns and impacts the uncertainty of the potential return on investment.

I will utilize these metrics, as well as the traditional LCOE metric, in the following sections to better characterize the potential economic viability and competitiveness of SMRs.

6.3 Affordability

An economic metric similar to LCOE is the "overnight capital cost." This term is a bit confusing because it is usually presented not as an absolute cost but rather as a normalized cost, that is, the total plant construction cost (not including financing cost) normalized by the rated output capacity of the plant and is typically expressed in units of dollars per kilowatt ($/kW). This is a useful parameter for comparing various plant options in the same market, but for many customers, the absolute plant cost is the dominant consideration and may be the showstopper. This is especially true for capital-intensive nuclear plants, which now appear to cost on the order of $6 to 8 billion for a large single-unit plant. To make matters worse, the most popular new nuclear plant design in the US is being sold as dual units, which effectively doubles the total cost to a teeth-jarring $14 billion. This price tag is prohibitively high for many potential owners. For these want-to-be customers, the cost per kilowatt is irrelevant if they cannot afford to purchase the plant.

When I decide to buy a new car and evaluate what I can afford, I think first about the sticker price and its implication on the monthly payment amount; next I evaluate the reliability reputation of the car as it impacts potential maintenance/repair costs; and finally, although only during times of high gasoline prices, do I worry about its fuel efficiency. If the sticker price of the car results in a monthly payment that is an unacceptably large fraction of my monthly income, I have to conclude that I cannot afford it. The bank helps me with this decision by refusing to loan me the money if I exceed their payment-to-income ratio guidelines.

While not a perfect analogy with buying a nuclear power plant, there are some similarities. Utilities have limits on what they can responsibly finance and what their investors or public service commissions will allow. The high cost of current large nuclear plants and the fragmented utility structure in the US are driving a situation in which only the few most affluent utilities can consider building new nuclear plants. This effect is portrayed in Figure 6.1, which compares the capital cost of four different energy sources: a multimodule SMR, a large dual-unit nuclear plant, a natural gas plant, and a coal plant. Cost data for the latter three plants were taken from an Energy Information Administration report.[7] The cost range for the SMR was gleaned from publicly available data for some near-term SMRs that are under development in the US. Figure 6.1 also shows the average annual revenue of all US investor-owned

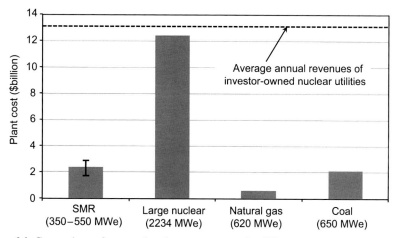

Figure 6.1 Comparison of power plant investment costs for different energy sources.

nuclear utilities.[5] The reality is that the purchase of a dual-unit large nuclear plant, similar to the plants being built in Georgia and South Carolina, is a "bet the company" proposition for the larger utilities and is beyond consideration for smaller utilities. Although producing significantly less power than a large nuclear plant, an SMR offers a sticker price that is within reach of the smaller utility and may better match their capacity needs anyway.

Another aspect of affordability is the flexibility to purchase an item in stages. There are many examples of this. A big selling point for some computers is the ability to upgrade the initial system at a later time with additional hard disks, more powerful graphics, or even new computer processors as the technology improves. Today's massively parallel supercomputers can be purchased "by the rack" and scaled to desired cost or performance goals. Houses are another example. For my former home, my wife and I initially bought a size of house that we could afford at the time, then added new space a few years later as our family grew. Multimodule SMRs offer this same benefit, which can have a significant impact on both the affordability and even the total plant cost due to savings in financing costs.

A group at the Politecnico di Milano (POLIMI) has been very active since 2007 in the economic assessment of SMRs. Led by my good friend Ricotti, they were a major partner in the multinational consortium developing the International Reactor Innovative and Secure (IRIS) SMR design and became the lead organization in 2010 when Westinghouse withdrew from the consortium. They were also a major contributor to the 2007–2008 IAEA study on economic competitiveness of small and medium reactors,[8] which was one of the first contemporary studies that looked specifically at the economic competitiveness of smaller-sized reactors. The study catalyzed POLIMI's development of the "integrated model for the competitiveness assessment of SMRs—the INCAS code." Subsequently, Ricotti, Boarin, Locatelli, and other team members have produced a number of insightful papers on various aspects of SMR economics, including the impacts of staggered deployment, sensitivity of investment to changes in electricity market conditions, and impacts of construction cost and schedule increases.[9–11]

Figure 6.2 shows an example of the impact of a staggered deployment of an SMR using INCAS-generated data provided by POLIMI.[12] I present three cases: (1) a compressed build-out of four 300 MWe single-unit SMR plants with a 1-year overlap of each consecutive 3-year SMR construction, (2) a protracted build-out of the same four SMRs with a 2-year gap between consecutive plant constructions, and (3) a single 1200 MWe plant with a 5-year construction duration. The construction schedules are depicted in the upper portion of Figure 6.2. In the lower portion of the figure, the cumulative cash outlay profiles for the two SMR cases are compared to the LR cash outlay. The maximum cash outlay for the LR reactor occurs in year five and is approximately 3.8 B€, which is the full amount of the plant's construction cost. For the compressed SMR build-out case, the maximum cash outlay is reduced by roughly 30% to 2.7 B€, which occurs in year eight. The reduction in peak cash outlay is a consequence of the fact that revenue is being generated by the earlier SMRs while the latter units are being constructed, thus increasing the amount of self-financing. This effect is even more apparent in the protracted build-out case, which results in a much smaller and

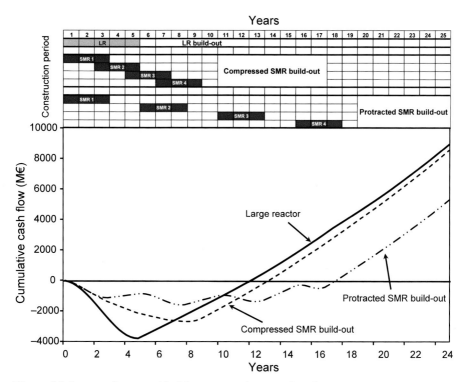

Figure 6.2 Impact of staggered build-out on maximum cash outlay.

flatter cash outlay profile. In this case, the maximum outlay is only 1.6 B€—nearly 60% lower than the LR case.

It is important to note that the three cases in this example result in different rates of new capacity addition to the grid. In the LR case, the full 1200 MWe of generation is available at the beginning of year six, while this does not occur with the SMR cases until year 10 for the compressed schedule and year 19 for the protracted schedule. However, if the reduced rate of cumulative capacity is acceptable to the plant owner based on demand growth, then the added flexibility of staggering the construction of the individual SMR units can have a significant impact on the owner's maximum debt. This in turn can help to reduce the overall project cost by reducing the cost of financing the plant's construction.

In assessing the total cost of a large construction project such as a nuclear power plant, it is essential to consider financing costs, referred to "interest during construction" (IDC). Obviously, the IDC is heavily influenced by the interest rate, which generally runs higher in open electricity markets, but it is also quite sensitive to the construction duration. For example, the IDC for a 4-year construction with an interest rate of 5% will represent roughly 15% of the overnight capital cost and nearly 30% if the construction duration is stretched to 10 years. In markets where the interest rate is 10%, the IDC contribution to the total project cost jumps to 30% for a 4-year construction and 75% for a 10-year construction time. Hence a major strategy for SMRs is to keep the total project cost as low as possible by implementing design features and plant architectures that enable more rapid and reliable construction schedules.

There are many other aspects of SMRs that contribute to reducing their cost, both capital and operational, and hence make them more affordable. Because most of these factors also contribute to the economic competitiveness of SMRs, I defer discussion of them to the next section. But this distinction is somewhat arbitrary—anything that reduces the cost of a nuclear plant will improve its affordability. However, I stress again that the economic competitiveness of a power plant with a very high total cost is only relevant to customers who can afford the plant's sticker price. SMRs are the most valuable to those who cannot.

6.4   Economic competitiveness

The economic competitiveness of SMRs continues to be their most hotly contested feature. The controversy is probably a result of many factors, foremost of which is an ingrained mindset in the nuclear industry favoring economies of scale. As discussed in a previous section, LCOE quickly became the dominant metric in the early phases of the nuclear industry. It was also widely assumed that LCOE is lowest for large plants due to economies of scale. I assert, however, that the fixation on the economies of scale principle actually drove the industry in the opposite direction, that is, it generated a succession of larger and more complex plants that ultimately resulted in a higher unit cost of electricity.

If we assume that LCOE is the best metric for judging the economic competitiveness of various options, the next issue becomes how best to achieve a low LCOE. Since LCOE is the total plant cost divided by the plant output, we have a choice of decreasing the numerator (cost), increasing the denominator (power), or both. I have had a long-standing debate on this issue with a colleague who works with a different reactor vendor. His belief, bolstered by the economies-of-scale mindset, is that the lowest LCOE will be achieved by designing the system to produce the maximum power possible within a given class of reactor size, i.e., increase the denominator. Unfortunately, this tends to lead to a design with high power density, forced circulation of the coolant, and complex safety systems—even for a relatively small reactor size. I believe that we should decrease the numerator and that system simplicity has the greatest impact on reducing total cost. Additionally increasing power is fine when possible, but not at the expense of simplicity.

6.4.1   Mitigating economies of scale

I freely admit that economies of scale are alive and well. Every system design must contend with this reality. However, a closer look at the economics of SMRs suggests that economies of scale can be substantially mitigated by a combination of other economic considerations. One of the earliest studies of the "economies of small" was conducted by Hayns and Shepherd of AEA Technology during the time of the development of the Safe Integral Reactor design.[13] They provided a qualitative discussion of a dozen factors that can offset the economies of scale in SMRs, which I summarize and paraphrase in Table 6.1.

Table 6.1 **Factors that can mitigate economies of scale in an SMR[13]**

Factor	Description
Increased factory fabrication	Smaller unit size permits a greater degree of factory fabrication and less on-site construction to reduce labor cost
Skilled workforce retention	Increased factory fabrication will help to attract and maintain skilled workers without the drawbacks of frequent relocations
More replication	Production of a larger number of small standardized units increases the efficiency of fabrication/construction
Multiple units at a single site	Shared infrastructure at a common site can reduce the capital cost per installed unit
Improved availability	Multireactor plant using smaller modular designs improves the possibilities for partial operation of the plant during maintenance of modules
Learning curve	Larger number of small identical units expedites achieving fabrication/construction efficiencies through learning process
Bulk ordering	Larger number of small units can reduce cost of bulk orders for individual components
Better match to demand	Small unit size allows better match to demand, thus reducing costs associated with purchased power or stranded assets
Smaller front-end investment	Smaller total project cost reduces the construction financing cost
Reduced construction time	Shorter construction time reduces the construction financing cost and introduces revenue sooner
Increased plant lifetime	Small modules may lead to easier refurbishment/replacement costs and extend lifetime of the facility
Reduction in planning margin	Smaller unit capacity shortens planning horizon and reduces investment sensitivities to market changes
Design appropriate to the size	Small unit size can enable simplification of systems and reduce costs of materials, fabrication, maintenance, and disposal

A study by Carelli et al. considered many of the same factors as well as a few additional ones.[14] They grouped the factors into two broad categories:

1. *Factors that are independent of plant size but may be enhanced by smaller units.* These include such things as modularization, factory fabrication, shared site infrastructure, and process learning.
2. *Factors that are unique to plants that are deliberately small.* These include such factors as design simplifications, plant compactness, demand matching, and economies of replication.

Carelli et al. quantified four of the factors for a specific case study that compared one 1340 MWe large plant against a plant comprised of four 335 MWe units (IRIS power level). In their study, each of the factors was assessed by evaluating actual data from the nuclear and other surrogate industries. Table 6.2 summarizes their findings. The column labeled "Individual" gives the value of the 4-unit SMR plant cost relative to the LR cost for the specific factor, while the "Cumulative" column gives the

Table 6.2 **Factors that can offset economies of scale for SMRs**[14]

Economic factor	Capital cost ratio (SMR/LR)	
	Individual	Cumulative
Economies of scale	1.70	1.70
Cositing of multiple units	0.86	1.46
Process learning	0.92	1.34
Construction schedule, timing, etc.	0.94	1.26
Modularity and design solution	0.83	1.05

progressive multiplication of the individual factors. Their results indicate that the 70% economies of scale penalty for the SMR plant is reduced to a nominal 5% penalty when considering the accumulation of the four offsetting factors. A few other studies have considered additional factors not quantified in the Carelli study, but to date, I have seen no comprehensive analysis of all factors.

Many economic studies have focused on "process learning" and "serial production (mass manufacturing)" as the dominant factors for mitigation of economies of scale in SMRs. Indeed, these factors are likely to be significant contributors to reducing SMR costs. The smaller component sizes of an SMR allow a greater percentage of the plant to be prefabricated in a factory, and their lower power output requires more units to be produced to generate the same total capacity as a large plant. Actual gains from these two factors vary among studies but generally range between 20 and 40%. As an example, Mitenkov observed the benefits of serial production in the range of 30–35% based on a study of serially produced propulsion units in Russia.[15] Because of the relative abundance of data for these factors from existing relevant industries, the study of process learning and serial production factors can be conducted with a higher level of confidence. Therefore, reliable models exist, and economists can rest easy in the accuracy and precision of their prognoses. While encouraging predictions have resulted from those studies, they have also generated some concerns as an unintended consequence of focusing on only process learning and mass production.

My first concern is that the nuclear industry has a poor record for demonstrating any benefit from process learning. There is ample evidence showing that the cost of nuclear plant construction in the US rose sharply with time despite increasing plant size. Arnulf Grubler of the Austrian International Institute for Applied Systems Analysis presents evidence that even for the relatively standardized build-out of the French nuclear fleet, there was no apparent savings from learning. Grubler asserts the following:

> *Lastly, the French nuclear case has also demonstrated the limits of the learning paradigm: the assumption that costs invariably decrease with accumulated technology deployment. The French example serves as a useful reminder of the limits of the generalizability of simplistic learning/experience curve models. Not only do nuclear reactors across the countries with significant programs invariably exhibit negative learning, i.e. cost increase rather than decline, but the pattern is also quite variable, defying approximations by simple learning-curve models...*[16]

This may be because true learning only applies if the same task is repeated by the same person in the same location. The protracted and diverse nature of nuclear projects may substantially mask the beneficial gains of learning. SMR vendors would be wise to remember this experience and not hang their future success on learning curves alone.

My second concern is that process learning is not a major differentiator among technology options and therefore has little to offer in evaluating the overall value proposition of a new technology. All technologies follow a similar learning pattern, that is, a significant reduction in cost for the first few units followed by a rapidly diminishing benefit for additional units. From the studies that I have observed, the cost reduction for the nth-of-a-kind unit flattens out at roughly 70% of the first-of-a-kind unit and occurs after construction of roughly 8–10 units. To reduce costs further, many more units are required.

This leads me to the third and greatest concern: focusing only on learning and serial production factors takes us down a path of logic that results in the conclusion that hundreds—maybe even thousands—of SMR sales are needed to make SMRs competitive. I have heard this argument offered by well-intended advocates who have not fully appreciated the logical conclusions. However, SMR opponents have been quick to seize this argument and extend it to conclude that SMRs are not worth pursuing until thousands of potential orders are assured. This is a hugely incorrect assumption, which has never been the case for the introduction of any new product. The appeal of SMRs will be determined in the success of the first few plants, and if achieved, many orders will follow that will further reduce costs and improve their appeal.

6.4.2   Enhancing economies of small

I offer a different view on the potential for SMR competitiveness: I believe that cost savings from design simplifications will end up being the dominant driver for those SMRs that are successfully competitive. My argument is based on the realization that "process learning" implies building the same item repeatedly to gain efficiencies. In contrast, "enlightenment" implies building a different item that provides similar functionality in a smarter way. An example that comes to mind, for those of you old enough to remember them, is the videocassette recorder (VCR). My very first VCR weighed more than five pounds and cost more than I care to admit. When I opened up the unit to see how it worked, I was amazed at the complexity of gears, levers, and switches. The last VCR that I bought weighed less than one pound and cost one-tenth the amount of my first one. Peering under the cover of this unit, I was equally amazed at how simple the mechanisms had become. Learning or enlightenment?

The primary basis for my assertion that design simplification will play the dominant role in SMR affordability and competitiveness is that it has a multiplying effect throughout the entire life cycle of the plant. Every component that is avoided in the design avoids costs associated with an engineer's time to design it, an analyst's time to evaluate its safety and performance, a regulator's time to review it, its purchase from a supplier, its installation, its maintenance, and ultimately its disposal at the end of the plant's life. Significant simplification—the kind that can only be achieved in smaller systems—should yield significant reductions in the plant's life cycle cost. There is one caveat to this argument, however. In the world of nuclear plant design, the regulator must carefully review every aspect of the design, including features that have been eliminated, to ensure that the absence of the system does not create a safety vulnerability. Hence, regulatory acceptance of the first occurrence of a simplified design may actually require more review effort, depending on the strength of the designer's case for the overall safety of the simplified design. Once accepted, however, subsequent licensing should be faster and cheaper.

A few of the economic studies cited earlier have hinted at the importance of design simplicity, but to date, none have performed any rigorous analysis of this factor. To be fair, it is a tougher job than analyzing learning curves. Access to the needed design data can be challenging due to the proprietary nature of detailed design information for SMRs that are under development. Also, there is considerable diversity of SMR designs, ranging from highly innovative designs to downsized versions of large plant designs. Because of this diversity, an objective analysis of cost impacts from design simplification could be valuable in comparing alternative designs.

Hayns and Shepherd recognized the importance of design simplification in their 1991 study:

> *A fundamental point which is sometimes missed is the fact that large plants have been optimized for their particular power output and it does not necessarily make sense in designing a smaller output plant to just scale down a larger system. At the smaller size a different design concept might be possible and using a design concept more appropriate for the reduced size—indeed one which is perhaps only technologically possible at the reduced size—could well lead to capital costs significantly lower than simple application of the scaling laws to the large design would predict.[13]*

They went on to discuss the 12 cost-reducing factors listed in Table 6.1 in the context of the Safe Integral Reactor design. A quantitative accounting of costs was conducted in Carelli's 2010 study and resulted in the 17% cost reduction due to design modularity and simplification for the IRIS design, as listed in Table 6.2. From my knowledge of several SMR designs being developed, this percentage probably represents a reasonable average with some designs having only minor simplification from large plants and others having substantially more simplification.

Similarly, Li recognized the importance of design simplification in his 2009 study.[17] He cites the example of emergency core cooling systems, which are added to a nuclear plant to protect against the severe consequences of a loss-of-coolant accident. He argues that the lower power rating of an SMR results in design solutions that allow simple passive systems to suffice. In contrast, large (high power) plants require sophisticated active emergency core cooling systems. In the case of the NuScale design, the unique integrated module configuration and small unit power—just 50 MWe—allows 15 major systems or components to be eliminated while introducing only one new system that is not found in traditional pressurized water reactors.

Although not in the category of design simplification, that is, removal of unneeded components, the smallness of SMRs facilitate additional cost savings related to the supply chain. As an example, the integral reactor vessel in most SMRs is sufficiently small to allow multiple manufactures, even domestic manufacturers, to create the vessel forgings. The limited number of global manufacturers that can produce the forgings needed for large nuclear plants has created a supply chain bottleneck that greatly impacts the cost and schedule for the fabrication of reactor and steam generator vessels. Until very recently, vendors had to pay substantial fees just to secure a spot in the job queue of a heavy component manufacturing facility such as Japan Steel Works.

As another example, the smaller turbine and generator equipment for SMRs can be sufficiently small that "off the shelf" options are available from multiple suppliers, resulting in more competition among suppliers and lower component costs. In addition, the smaller equipment is more easily transported to the site and removed for repair or replacement. In an article on safety considerations of hydrogen-cooled generators, Spring of *Power Engineering* magazine indicated that over 70% of all generators greater than 60 MW use hydrogen cooling.[18] In addition to requiring rigorous transport and handling procedures, the hydrogen cooling system complicates generator maintenance. In contrast, NuScale's 50 MWe turbine/generator equipment is sufficiently small to be shippable on a single pallet, and the generator can be air-cooled rather than hydrogen-cooled.

6.4.3 Diseconomies of scale

A number of studies have emerged that explore the diseconomies of scale, that is, the potential for costs to increase with increasing unit size. In a study of the high construction costs for nuclear projects, Kessides of the World Bank concluded that a significant factor was the misjudging of the economies of scale: "early projections tended to ignore the potential diseconomies of scale due to the increased complexity and greater management requirements of larger nuclear plants."[19] An analysis specific to SMRs was conducted by Li of Intellectual Ventures, who concluded that there is no evidence that the nuclear industry has observed any economies of scale despite the rapid scale-up in plant size over the past several decades.[17] While some would like to attribute the cost escalation of nuclear plants to regulatory or other factors unique to the nuclear industry, Li presents evidence that similar diseconomies of scale appeared much earlier in the coal industry.

A 1979 study by Ford at the Los Alamos Scientific Laboratory (now the Los Alamos National Laboratory) supports Li's observation. Ford conducted a comparative analysis of building one 3000 MW coal plant versus six 500 MW plants.[20] He concluded that the collective set of small coal plants was nominally 5% more economical than the single large plant, even with the small plants sited at different locations, that is, not benefiting from shared site infrastructure. Ford's report cited many of the same cost-saving factors cited by Hayns and Carelli in their studies. The dominant factor in favor of smaller coal plants was maintenance cost, including the cost of makeup power during the more frequent maintenance outages of the large, complex coal plants. He also cited a number of nonfinancial benefits of smaller units. Benefits specific to coal plants included air quality improvements resulting from the distributed siting of the plants and visual aesthetics due to the reduced height of the smoke stacks. Benefits that apply to SMRs as well included such factors as lower water usage, reduced demand forecasting requirements, and reduced "boom town" impacts.

Interestingly, at about the same time as Ford's study, new coal plant construction moved toward smaller unit sizes, eventually settling in on nominally 200 MW as a manageable size for a single generating unit. As an example, the massive Kingston Steam Plant near my home in Tennessee produces roughly 1400 MWe from nine

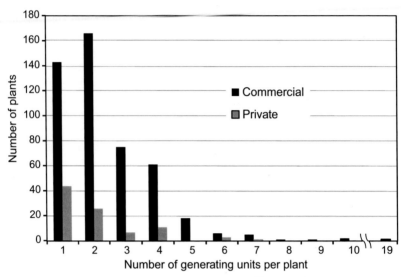

Figure 6.3 Distribution of multiunit coal plants in the US.[21]

smaller generating units of 175–200 MWe each. Figure 6.3 shows the number of coal-fired generating units per plant for both commercial and private coal plants in the US.[21] The most modular of the plants is owned by the Eastman Chemical Company in Tennessee, which operates 19 small coal-fired boilers.

Another supportive study of the diseconomies of scale was published by Dahlgren and colleagues at Columbia University.[22] Their paper, "Small Modular Infrastructure," provides a discussion of SMRs, although they also explored small modular chlorine plants, biomass gasification plants, and other similar technologies. They provide a rigorous analysis of several cost-impacting factors, including both capital and operating costs, and offer alternative interpretations of some of the relationships that have been assumed to support economies of scale. A clear example of this is their analysis of the cost of commodities. Traditional wisdom is that the utility of a component scales by its volume whereas the amount of material, and hence the commodity cost, scales by its surface area. As a component increases in size, its volume increases faster than its surface area—hence an economy of scale. The authors contend that structural mechanics considerations will tend to reduce the advantages of larger units and ultimately increase cost as the component size is increased. The authors conclude the following:

> It is now realistic to consider a radically new approach to infrastructure design, one that replaces economies of unit scale with economies of numbers, that phases out custom-built, large-scale installations and replaces them with large numbers of mass-produced, modular, small-unit-scale technology.[22]

Before moving on to other economic considerations, I would like to share my favorite analogy for the diseconomies of scale. It comes from a 2012 presentation by Locatelli at the University of Lincoln in the United Kingdom.[23] In his presentation,

Figure 6.4 Example of diseconomies of scale: a single ostrich egg versus a dozen chicken eggs.

he compared an ostrich egg to a dozen chicken eggs, as shown in Figure 6.4. The going price for an ostrich egg in 2012 was 15–25 € for a single egg weighing 1.5–1.8 kg, yielding a unit price of 10–15 €/kg. In contrast, a dozen chicken eggs sold for 2–4 € and weighed 60–70 g per egg, yielding a unit cost of only 2.5–5 €/kg or roughly one-third the unit cost of the single larger egg. The higher unit cost of the ostrich egg is attributed to the fact that it is a relatively unfamiliar product to most people and is offered by only a few uncommon suppliers. On the other hand, chicken eggs are quite familiar and available from an abundance of common suppliers. In addition to the cost disadvantage, the ostrich egg feeds a family of 15–20 people, generally not a good match with most family sizes.

6.4.4   Other economic considerations

Most of this chapter has dealt with capital cost, that is, the cost of the physical plant. Operations and maintenance (O&M) costs are also important and key to the competitiveness of SMRs. Unfortunately, there is a higher level of uncertainty in SMR O&M costs than their capital cost because of the lack of operational experience with SMRs. There are also uncertainties associated with the regulator's acceptance of key differences from traditional plants, especially for multimodule SMRs. For instance, most SMRs are being designed with all of their safety systems located below ground level to reduce their vulnerability to airborne attack and to limit access pathways to critical systems. Designing physical protection into the intrinsic features of the plant would appear to be the preferred approach compared to the classic strategy of more "guns, guards, and gates." But we will not know for a few years if the regulator will allow plant vendors to receive credit for these intrinsic features in terms of a reduced size of the security force. This uncertain outcome will have a significant impact on the plant's O&M costs and its competitiveness.

A similar uncertainty exists for control room staffing requirements. The current NuScale design, of which I am particularly familiar, includes a single control room for the operation of up to 12 modules using fewer reactor operators than would be required to operate 12 LR units. This poses a regulatory uncertainty that has direct implications on O&M costs. Other unique aspects of SMRs that will impact their relative O&M costs include factors such as refueling strategy (dedicated staff vs temporary workforce), in-service inspection requirements, the level of system automation, fleet-wide management of highly standardized modules, and fleet-wide spare parts management.

There are several other potential benefits of SMRs that have obvious economic impact but are challenging to monetize. In 2011, Locatelli and Mancini at POLIMI published a thorough paper on several nonfinancial factors related to plant size that can impact investment decisions.[24] They identified 11 classes of considerations: spinning reserves management, electric grid vulnerability (stability), public acceptance, technical siting constraints, project risks, national industrial system, time to market, competencies required for operation, employment impacts, design robustness, and historical/political relationships. Although their analysis was mostly qualitative, smaller-sized plants appear to have the advantage for nearly every factor considered. One of the few factors that may tend to favor large plants is the historical/political relationships factor, that is, a previously established relationship between a vendor and a customer or their respective countries. This will apply to siting of new plants in countries or regions that have an established relationship with a nuclear supplier who only offers large plant designs.

A final point that I want to make on economic competitiveness is that throughout the preceding discussion, we have never discussed the important question of "competitive with what?" This is a major ambiguity associated with the competitiveness dialogue. Too often, the assumption is that SMRs must be competitive with large nuclear plants. Going back to the opening discussions in this chapter, SMRs are primarily for the benefit of customers who cannot afford, or do not need, large nuclear plants. Many of these customers are not situated on large-capacity grids, and their alternative energy choices may be more constrained and expensive. Some studies are beginning to recognize this distinction, such as the study by the Institute for Energy and Transport at the European Commission's Joint Research Center.[25] In this study, Carlsson and colleagues specifically evaluate the competitiveness of SMRs for European cogeneration markets, which appears to be more favorable than for European electricity markets as evaluated in a related study led by Shropshire.[26] A study led by Harrison at Oak Ridge National Laboratory also recognized this distinction. They observed that the smaller power output from SMRs allows them to compete very favorably outside the large wholesale markets, such as for industrial facilities and military installations.[27] Indeed, SMRs are intrinsically well suited for meeting localized electrical demand and nonelectrical applications. I defer the discussion of these important opportunities for SMRs until the next chapter.

6.5   Reducing economic risk

With respect to affordability, total project cost is clearly the dominant factor; however, financing options are a close second. In the case of my earlier analogy of buying a car, the bank has a lot to say about the age and cost of the car I buy. While part of their concern is my ability to make payments, they also care more broadly about any potential risk to their investment. The bigger the risk, the higher their concern and the more they charge me for using their money. As risk increases, some financing options disappear.

Undeniably, nuclear power conjures up a higher perception of risk to the general public than most other industrial endeavors. Maybe it is the lingering memories of cold-war threats of nuclear annihilation, maybe it is the mystical radiation that accompanies nuclear power, or maybe it is the way we package it in large, remote facilities. This broad perception of risk permeates even into the investor community, although for reasons that are less emotional and more founded on financial principles. I got my first glimpse of investors' views of nuclear risks when I helped to plan and host a 2009 workshop on financing of nuclear projects as part of the Global Nuclear Energy Partnership (GNEP) program. In addition to the usual technical representatives from the nuclear industry, several individuals were invited from the financing community—a community that is often overlooked in these types of meetings. Representatives were present from Deutsche Bank, Fitch Rating Services, Towers Perrin, Nippon Export and Investment Insurance, and the US Export-Import Bank. These representatives shared a consistent message: nuclear projects are among the most difficult for investors to embrace. Some of the more common justifications included the capital-intensive nature of nuclear projects with very long payback times, the impact of external factors such as regulatory uncertainties and public response, construction delays and cost overruns, and the need for reliable long-term market forecasts. In Grubler's paper on the French nuclear program, he writes the following: "Perhaps the nuclear 'valley of death' is its inherently high investment costs and their tendency to rise beyond economically viable levels."[16]

In Chapter 10 of the *Handbook of Small Modular Nuclear Reactors*, the POLIMI group discusses financing risks associated with nuclear power projects.[4] They cite many of the same issues that were voiced by the bankers at our GNEP meeting, which they categorize into those that are associated with capital-intensive projects and those that are unique to nuclear projects. Their major risk factors are delineated in Table 6.3.

The POLIMI group further asserts that construction cost and schedule overrun are the leading factors that contribute to nuclear project risk. As mentioned earlier, total project cost is highly sensitive to construction duration. The most significant component of this sensitivity is the financing cost, but other cost factors can contribute such as retention of the labor force and protracted rental fees for major construction equipment. POLIMI's paper includes a painful example of historical data that helped to earn the poor reputation of nuclear power projects. According to data obtained from

Table 6.3 Major risk factors impacting financing of nuclear power plant projects[4]

Common to capital-intensive projects	Unique to nuclear power projects
• High up-front capital costs • Cost uncertainty • Construction supply chain risks • Long lead times and long payback periods • High sensitivity to interest rates • Plant reliability and capacity factor • Market price of product (electricity)	• Unstable public support • Negative public acceptance • Regulatory and policy changes • Decommissioning/waste disposal costs and liabilities

the US Congressional Budget Office,[28] the actual cost of the 75 nuclear plants initiated in the US between 1966 and 1977 consistently ranged between two to three times higher than their original cost estimate. The inherent features of SMRs mitigate many of the risks, most of which have been discussed in previous sections. These include a shorter construction time, faster time to first revenue, and higher reliability in fabrication costs and schedule. The shorter construction time and sequential construction features also result in lower sensitivity to several market-related factors such as interest rates, electricity prices, and changes in electricity demand due to the owner's ability to respond faster to changing conditions.

During the 2009 GNEP workshop mentioned earlier, the banker's presentations were followed by presentations by SMR vendors, who described the features of their designs and deployment strategies. Throughout these latter presentations, you could visibly see a change in the attitude of the bankers—they became more engaged, more inquisitive, and even smiled occasionally. In the wrap-up session, the financing representatives acknowledged that SMRs appear to reduce many of the major investment risks associated with nuclear projects, most notably because they require less total capital investment per module, can be constructed more quickly and reliably (in-factory vs on-site), and can be generating revenue earlier to offset the cost of subsequent modules. Of special note was the fact that the combined effect of a shorter construction time and reduced capacity of an SMR helps to mitigate the impact of miscalculating the long-term forecast of electricity demand, which can result in the purchase of expensive replacement power or a stranded asset—either of which hurts the plant's profitability.

While SMRs appear to have a clear advantage over single large plants in terms of reducing investment risks listed in the left-hand column of Table 6.3, it is less obvious that they will have any impact on the nuclear-specific risks. Their ability to impact these factors will depend substantially on the successful licensing and deployment of the first SMRs in terms of demonstrating the first two major strengths of SMRs: their enhanced safety/robustness and their affordability. If this can be done for traditional electricity applications, it will enable the realization of the third major benefit of SMRs: their suitability for expanding the use of nuclear power to nontraditional markets and nonelectrical applications—the topic of the next chapter.

References

1. Rowe JW. My last nuclear speech. In: *Presented at the American nuclear society utility working conference, Hollywood, FL.* August 2011.
2. Weinberg AM, Spiewak I, Barkenbus JN, Livingston RS, Phung DL. *The second nuclear era.* Praeger Publishers; 1985.
3. Locatelli G, Bingham C, Mancini M. Small modular reactors: a comprehensive overview of their economics and strategic aspects. *Prog Nucl Energy* 2014;**73**:75–85.
4. Boarin S, Mancini M, Ricotti M, Locatelli G. Economics and financing of small modular reactors. [Chapter 10]. In: *Handbook of small modular nuclear reactors.* Cambridge, UK: Woodhead Publishing; 2014.
5. Rosner R, Goldberg S. *Small modular reactors—key to future nuclear energy power generation in the U.S.* Energy Policy Institute at Chicago, University of Chicago; November 2011.

6. *Current status, technical feasibility and economics of small nuclear reactors.* Nuclear Energy Agency; 2011.
7. *Updated capital cost estimates for utility scale electricity generating plants.* U.S. Energy Information Administration; April 2013.
8. *Approaches for assessing the economic competitiveness of small and medium-sized reactor.* International Atomic Energy Agency; 2013. NP-T-3.7.
9. Boarin S, Ricotti M. Cost and profitability analysis of modular SMRs in different deployment options. In: *Proceedings of the 17th international conference on nuclear engineering (ICONE), Brussels, Belgium, July 12–16, 2009.*
10. Boarin S, Locatelli G, Mancini M, Ricotti M. Financial case studies on small- and medium-sized modular reactors. *Nucl Technol* May 2012;**178**:218–32.
11. Boarin S, Ricotti M. An evaluation of SMR economic attractiveness. *Sci Technol Nucl Install* 2014;**2014**. Hindawi Publishing Corporation.
12. Boarin S. Private communication, January 11, 2015.
13. Hayns MR, Shepherd J. SIR: reducing size can reduce cost. *Nucl Energy* April 1991;**30**(2):85–93.
14. Carelli MD, et al. Economic features of integral, modular, small-to-medium size reactors. *Prog Nucl Energy* 2010;**52**:403–14.
15. Mitenkov FM, Averbakh BA, Antyufeeva IN. Economic effect of the development and operation of serially produced propulsion nuclear power systems. *At Energy* 2007;**102**(1):42–7.
16. Grubler A. The costs of the French nuclear scale-up: a case of negative learning by doing. *Energy Policy* 2010;**38**:5174–88.
17. Li N. A paradigm shift needed for nuclear reactors: from economies of unit scale to economies of production scale. In: *Proceedings of the international congress on advanced power plants (ICAPP), Tokyo, Japan, May 10–14, 2009.*
18. Spring N. Hydrogen cools well, but safety is crucial. *Power Eng* June 1, 2009. Available at: http://www.power-eng.com/articles/print/volume-113/issue-6/features/hydrogen-cools-well-but-safety-is-crucial.html.
19. Kessides IN. The future of the nuclear industry reconsidered: risks, uncertainties, and continued promise. *Energy Policy* 2012;**48**:185–208.
20. Ford A. *A new look at small power plants.* Los Alamos Scientific Lab; January 1979. LASL-78-101.
21. *Annual electric generator report, Form EIA-860.* Energy Information Administration. Available at: www.eia.gov/electricity/data/eia860.
22. Dahlgren E, et al. Small modular infrastructure. *Eng Econ* 2013;**58**(4):231–64.
23. Locatelli G, Mancini M, Ruiz F, Solana P. Using real options to evaluate the flexibility in the deployment of SMR. In: *Presented at the international congress on advances in nuclear power plants, Chicago, IL, June 24–28, 2012.*
24. Locatelli G, Mancini M. The role of the reactor size for an investment in the nuclear sector: an evaluation of non-financial parameters. *Prog Nucl Energy* 2011;**53**:212–22.
25. Carlsson J, et al. Economic viability of small nuclear reactors in future European cogeneration markets. *Energy Policy* 2012;**43**:396–406.
26. Shropshire D. Economic viability of small to medium-sized reactors deployed in future European energy markets. *Prog Nucl Energy* 2011;**53**:299–307.
27. Harrison TJ, Hale RE, Moses RJ. *Status report on modeling and analysis of small modular reactor economics.* Oak Ridge National Laboratory; March 2013. ORNL/TM-2013/138.
28. *Nuclear power's role in generating electricity.* US Congressional Budget Office; May 2008.

Expanding nuclear power flexibility

7

In this chapter, I complete my discussion of the three major small modular reactor (SMR) strengths: enhanced safety and robustness, improved affordability, and expanded flexibility. Safety is a go/no-go requirement that cannot be compromised if nuclear power is to remain an acceptable energy option. Safety alone is not sufficient for nuclear to be a viable contender in global energy markets—it must also be affordable and competitive. Having demonstrated the promising potential for safety and affordability of SMRs in the previous two chapters, I now turn to their flexibility of use. This third major strength of SMRs is central to expanding the use of nuclear power to a broader range of new customers and applications.

The first two features of SMRs that help to provide their expanded flexibility are captured in their name: small and modular. Two additional important features that are derived from their smallness and modularity are worth highlighting: their benefits for more flexible plant siting and their adaptability to nonelectrical applications. All four factors contribute to the promising potential that SMRs offer to expand the use of nuclear energy beyond the traditional application of baseload electricity generation in large-grid markets.

7.1 Size matters

It should not be a shocker that SMRs are small. I often get asked: How small can they be? I prefer to respond in terms of power output, which is a much easier response but still surprisingly complicated. There is general agreement that the upper bound on "small" is 300 MWe, although there is nothing magical about this precise number. However, the lower bound is quite fuzzy. The quick answer is that they can be as small as you want, all the way to zero. This is a true fact—many zero-power reactors have been built—but it is highly misleading. Zero-power reactors produce virtually no heat and are used only for research or to train students on the physics of critical nuclear assemblies. If you recall from the previous chapter, the definition of levelized cost of electricity (LCOE) is the total cost of the plant divided by its output power, so a zero-power reactor would have an LCOE of infinity—a remarkably poor investment for commercial applications.

Based on commercial investments in SMRs globally, it would appear that the lower practical limit for commercial use is on the order of several megawatts. Table 7.1 lists 20 SMRs that are currently being developed by commercial organizations.[1,2] Reactors for noncommercial use such as naval propulsion fit within the power range of these commercial designs. Reactors for space exploration tend to be much smaller, typically in the range of several kilowatts to as high as a few megawatts. So setting aside the issue of commercial viability, a reactor can be designed to whatever power is desired.

Small Modular Reactors. http://dx.doi.org/10.1016/B978-0-08-100252-0.00007-0

Table 7.1 **The power capacity and reactor vessel size of several commercial SMRs under development worldwide**

Country	SMR design	Coolant	Power (MWe)	Reactor vessel size	
				Diameter (m)	Height (m)
Argentina	CAREM	Light water	27	3.2	11
China	ACP-100	Light water	100	3.2	10
China	CNP-300	Light water	300–340	3.7	10.7
China	HTR-PM	Helium	105	5.7	25
France	Flexblue	Light water	160	3.8	7.7
India	PHWR-220	Heavy water	235	6.0	4.2
India	AHWR-300-LEU	Light water	304	6.9	5.0
Japan	4S	Sodium	10 or 50	3.5	24
Rep of Korea	SMART	Light water	100	5.9	15.5
Russian Fed	ABV-6M	Light water	8.5	2.1	4.5
Russian Fed	KLT-40S	Light water	35	2.1	4.1
Russian Fed	RITM-200	Light water	50	3.3	8.5
Russian Fed	VBER-300	Light water	300	3.7	8.7
Russian Fed	SVBR-100	Lead-bismuth	101	4.5	7.9
US	mPower	Light water	180	3.9	25.3
US	NuScale	Light water	50	2.9	17.4
US	SMR-160	Light water	160	2.7	13.7
US	W-SMR	Light water	225	3.5	24.7
US	EM2	Helium	265	4.7	10.6
US	PRISM	Sodium	311	9.2	19.4

Here is a related question that frequently follows: How big (physically) are small reactors? My not-so-helpful response is typically "smaller than large reactors." The reason for my coyness is that there is considerable size variation among the various SMR designs, as shown in Table 7.1. The table lists the size of only the reactor vessel, which is proportionally smaller for a loop-type design than for an integral design. On the other hand, loop designs have multiple vessels outside of the reactor vessel, such as steam generator and pressurizer vessels, which are still part of the primary system and add to the overall dimensions of the plant. As demonstrated in the table, SMRs are not necessarily small from a physical size perspective. For example, the mPower integral SMR has a reactor vessel that is as tall as an 8-story building. Another feature that adds size to an SMR is the containment structure that surrounds the reactor vessel. The design strategy for the containment enclosure varies widely among SMR designs; some, such as NuScale, use a small-volume steel vessel, while others, such as SMART, use more traditional large-volume concrete structures. For comparison, more than 120 NuScale containment vessels could fit within a single containment building of a traditional large reactor.

The third facet of SMR "smallness," after electrical output and hardware size, is the plant's "footprint," that is, the amount of land used by the reactor building, all auxiliary buildings, and associated structures. This also varies among designs. For example, the Russian ABV-6M and KLT-40S reactors are sufficiently compact that an entire two-unit plant can fit on a single barge. For land-based SMRs, most vendors advertise a plant footprint ranging from 8000 to 16,000 m² (20–40 acres), which is roughly one-tenth the size of a large plant footprint. Many SMR plants have not only a small footprint but also a low building profile due to below-grade placement of safety systems, smaller-scale turbine equipment, and the use of forced draft cooling towers. Radioactive waste handling facilities, the administration building, the switchyard, and the dry cask storage area complete the SMR site. All three size considerations—energy output, hardware dimensions, and plant footprint—are important in understanding the potential deployment of SMRs to different markets and their suitability for diverse energy applications.

7.1.1   Remote customers

The first potential benefactors of SMRs are remote electricity consumers. There are practical limits to this, however. A widely publicized example is the town of Galena, Alaska. Beginning in the mid-2000s, Galena city officials pursued an arrangement with Toshiba to replace the city's diesel generator with an SMR, specifically the Super Safe, Small and Simple (4S) reactor. The notion of a simplified small reactor with the potential to provide 30 years of sustained power on a single fuel loading was very enticing since fuel to feed the town's diesel generator is very expensive and can only be resupplied during one month of the year. In the end, the project was canceled because even the 10 MWe output of the 4S plant was substantially higher than the 2 MWe peak load of the town. In another example, I was approached by a company bidding on a contract to upgrade infrastructure at the Nashville airport, a facility with an electrical load of roughly 0.5–1.0 MWe. The bidders were excited at the prospects of proposing to meet this demand with an SMR until I recalibrated their ambitions.

Although SMRs may not be a solution for the very remote and small consumer, there are many towns, cities, and facilities that have significant electrical demand but for which large nuclear plants or other energy alternatives may not be practical. Some of these communities are well positioned on adequate electrical grids, but many more have outgrown their local grid infrastructure or have only marginal grid interconnection. For these customers, SMRs offer a very promising solution.

I naively assumed that it was rather trivial (and inexpensive) to build new transmission lines. As it turns out, installing new overhead high-voltage lines can cost 1 to 2 million dollars per mile, and the cost of underground lines can be four to five times higher. In the western US, where I spend a lot of my time, it is especially challenging to add new transmission lines. Whether because of the vast amount of protected lands, the harsh topology, or safeguarding of endangered wildlife, the acquisition of new transmission right-of-ways and the construction of the lines are both difficult and expensive. I recently sat in on a meeting with some western utility representatives who were very animated about the challenges of building new transmission networks. One representative stated flatly that he would rather deal with constructing multiple SMRs in his service area than to fight the sage grouse lobby. This reality highlights a significant benefit of an SMR's smallness: its small unit output allows for a more distributed model of nuclear power generation than currently exists. To reiterate an earlier message, SMRs are about offering more choices to the utility executives, not only choices among power generation options but also the choice between distributed generating assets verses new transmission networks.

7.1.2 Grid management

Having smaller-sized generators helps to provide grid stability, especially in regions with limited grid interconnection. An often-cited rule of thumb for stable grid operation is that the output of a single generating unit should not exceed 10% of the total generating capacity. Even in large-grid regions, utilities are accustomed to operating smaller-sized generating units. Figure 7.1 shows the size distribution of all types of electrical generation plants worldwide, which indicates that roughly 93% of the generating plants have capacities below 500 MWe.[3] Even in the US, only 6 of the nearly 1400 new generating units added between 2000 and 2005 had power ratings of greater than 350 MWe. Consequently, most utilities are very familiar with the operation and grid management strategies for smaller generating units.

Related to this is a term called "spinning reserve," which is the excess capacity that must be available to the utility to accommodate the loss of the largest generator on the system without major disruption. Large, single-unit generating plants dictate the need for a large amount of spinning reserve. This can be accomplished by running many of the existing stations at slightly less than full capacity or by operating dedicated units on continuous standby. Either way, energy is being wasted. The smaller output of an SMR provides a more manageable power distribution and reduces wasted resources associated with the spinning reserve capacity.

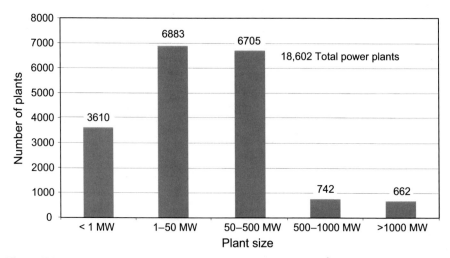

Figure 7.1 Size distribution of electricity generation plants globally.[3]

Also related to the integrity of the grid infrastructure is the debate about load-following capabilities of nuclear plants. Conventional wisdom suggests that nuclear power plants should be operated continuously at full capacity and that natural gas plants are best suited to provide "peaking" capability to meet excess demand. This historical strategy has been driven in part by economic considerations. The logic is that because a nuclear plant has a relatively high capital cost and relatively low fuel cost compared to a natural gas plant, it makes sense to keep the more expensive hardware running continuously. Because of its low fuel cost, running the nuclear plant at 50% power has a minimal impact on its operational costs but cuts revenue in half. Increased maintenance costs are also normally higher with load-following operation due to fatigue issues introduced by the thermal cycling of the reactor and balance-of-plant systems.

Despite the economic argument, many nuclear plants currently operating in the US were designed to load-follow and were originally outfitted with automatic grid control features. However, the US Nuclear Regulatory Commission established a policy that precluded the use of automatic dispatching of nuclear plants, although they allow manual load-following if conducted by a licensed reactor operator. Only the Columbia Generation Station in Richland, Washington performs some degree of load-following operations, which they refer to as load-shaping. Worldwide, France's pressurized water reactors load-follow on a regular basis due to the high percentage of nuclear-generated electricity on their grid (nominally 75%). Canadian reactor units are also required to load-follow due to the relatively high percentage of nuclear power there, and German reactors load-follow primarily because of a relatively high contribution of intermittent wind generation on their grid.[4]

There are now two reasons to consider load-following with nuclear plants. The first is for small grid applications enabled by SMRs. In these cases, similar to what is experienced in France and Canada, the relatively high percentage of nuclear generation on the regional grid will likely force the SMRs to follow the changing demand. The second reason to consider load-following with nuclear plants is in response to an increasing penetration of renewables on the grid, similar to the situation in Germany. The increasing penetration of renewable sources, especially wind, has altered the economic argument since wind turbines are also capital-intensive (more expensive than nuclear plants per unit of power produced), and their fuel cost is very low—zero, in fact. Also, some regional policies require grid dispatchers to preferentially use renewable energy first, thus requiring base-load plants to accommodate demand variations. Consequently, plant owners are already beginning to perform power maneuvering with coal plants and are being pressured to extend this to nuclear plants.

Fortunately, the smallness of a typical SMR allows it to better accommodate load-following operations, and several vendors advertise this capability. In general, load-following impacts are reduced relative to large plants because of the greater agility of the SMR to respond to thermal cycling. The greater agility is provided by the smaller reactor system components, smaller turbine/generator equipment, and system simplifications that are enabled by the smaller reactor size. Fast spectrum reactors such as sodium-cooled SMRs have a further advantage in that their core reactivity is much less sensitive to xenon buildup and depletion in the reactor core, a phenomenon that complicates core power maneuvers in thermal spectrum reactors such as water-cooled reactor systems.

Several SMR vendors advertise some degree of enhanced load-following capabilities. Drawing on my familiar example of the NuScale design, it incorporates specific features to enhance its ability to load-follow, either in response to changes in electricity demand or variable generation by renewable sources on the grid. This is accomplished through a combination of the small unit capacity of a NuScale module (50 MWe) and a multimodule approach to the plant design. One example is that the module design and operating parameters allow reactor power changes using only control rod movement down to 40% reactor power, that is, it does not require adjustments to the boron concentration in the primary coolant. This improves the maneuverability of the reactor while not creating additional liquid wastes associated with boron addition and dilution. Also, the condenser is designed to accommodate full steam bypass, thus allowing rapid changes to system output while minimizing the impact to the reactor, which can continue to run at full power. For more substantial power maneuvering, entire modules can be shut down, thus reducing the plant output in increments of 50 MWe. A study looked at integrating a NuScale plant with the Horse Butte wind farm in Idaho and found that the plant could effectively offset the variability of the wind farm using either turbine bypass alone or a combination of bypass and power maneuvering.[5] Still, load-following with a nuclear plant is not without consequences, including economic and mechanical implications. So while SMRs are likely to accommodate load-following better than large plants, it is not the preferred mode of operation.

7.1.3 Nonelectrical customers

Another operational flexibility with smaller nuclear plants is the ability to better match the plant output to the user requirements, which is especially valuable for process heat applications. I will defer discussion of the technical aspects of coupling SMRs to process heat facilities until later in this chapter and limit the discussion here to the size suitability of SMRs for nonelectrical applications. Somewhat surprising to me is that many process heat users actually require a relatively modest amount of heat—far less than the 3000–4000 MWth (1000–1400 MBtu/h) that a large nuclear plant produces. For example, a 2008 study by Sherrell Greene, a good friend and former boss at Oak Ridge National Laboratory (ORNL), concluded that a nuclear plant producing approximately 100 MWt is sufficient to provide all the process heat needed for the production of liquid fuels in a state-of-the-art biorefinery.[6] This modest heat demand is a result of the fact that the biorefinery size is dictated by the transportation cost associated with transport of the bulk biomass, which limits the reach of the feedstock supply to about 50 miles.

Most studies of process heat applications aggregate energy usage on a country or global basis. This is useful for understanding the magnitude of the potential market but does not speak to the options for delivering heat to individual sites. A few studies have been published that provide data on the energy demands of individual industrial plants. In Ning Li's paper entitled "A Paradigm Shift for Nuclear Reactors: From Economies of Unit Scale to Economies of Production Scale," he indicated that the thermal demand for a typical 200,000 barrels/day coking refinery is roughly 1100 MWth.[7] He further suggested that reliability requirements would imply modularity of the heat source, perhaps three 400–600 MWth modules. Li also observed that a typical alumina refinery requires roughly 100 MWth and 33 MWe. Similar results were reported in a NuScale study, which found that approximately 1700 MWth of externally generated heat are needed for a 250,000 barrels/day oil refinery.[8] A second NuScale study indicated that a commercial-scale water desalination plant producing 190,000 m³/day (50 million gallons per day) of clean water consumes between 150 and 550 MWth, depending on the choice of desalination technology.[9] Finally, a paper published by the European Commission's Joint Research Institute indicates that in the European industrial market, the average thermal demand per site of several industrial processes ranged between 100 and 400 MWth. This included factories for the production of various products such as chemicals, cement, iron and steel, soda ash, and alumina.[10]

In summary, size does matter. For regions with large electrical demands and adequate transmission infrastructure, large nuclear plants offer an attractive solution. For smaller regional demands and locations with limited or no grid interconnections, SMRs provide a unique solution for abundant, clean, and reliable energy. For industrial users, SMRs can be scaled to cover a wide gamut of thermal energy applications.

7.2   Benefits of modularity

Related to small size is the equally important factor of modularity. They work together to provide a scalable solution to meet a wide range of potential applications. As I take a quick look around the room, I see several examples of modularity. One example is a recent gift from my daughter: a pair of radio-controlled bumper cars (actually, quite fun to play). Each car and each controller requires multiple batteries for a grand total of 14. While this was a surprising number of batteries for just one toy, the power demand was met using a single type of battery: an AA battery. Fortunately, I had a convenient stock of these on hand for just such a purpose. The small, modular nature of the AA battery allowed the toy manufacturer to tailor the power requirements of the various components of the toy—four for each car and three for each controller—using a single standardized energy cell. As an added benefit, the batteries were relatively cheap due to mass manufacture.

A more sophisticated example is the supercomputer. When I began my career at ORNL, the mainframe computer was an IBM model 360. It was big and expensive, and only the most affluent research organizations had one. This trend in mainframes continued as the laboratory upgraded to bigger and faster single-processor computers, ultimately acquiring a Cray YMP supercomputer at a cost of several million dollars. Then in the early 1990s, IBM introduced the personal workstation that provided decent computing capability for a single user and cost a few thousand dollars. Researchers quickly discovered that these units could be clustered together to create departmental computing servers for a small fraction of the cost of the laboratory's mainframe. They also appreciated the added benefit of having full control over how the computing resource was managed rather than being at the mercy of a centralized Information Technology group. The move from single-processor mainframes to workstation clusters served to bridge the computing community to an entirely new computing paradigm: parallel computing. Today, ORNL's showcase supercomputer, designated

Titan, yields a phenomenal 20 petaflops (20 million billion calculations per second) of computing power using 299,000 small processors in parallel.

In addition to the familiar case of consumer batteries, examples of modularity exist throughout the energy industry. The world's largest solar farm, in California, produces over 500 MWe from nine million individual solar panels. Multigigawatt wind farms are built from thousands of 1–2 MWe wind turbines. Also, I mentioned in the previous chapter that many coal plants are comprised of multiple small boilers such as the 1400 MWe Kingston Steam Plant in Tennessee that uses nine 175–200 MWe boilers. In fact, nuclear appears to be the last holdout in the energy industry for the sole use of large single-unit power generating stations.

Multimodule plants allow the owner to build new capacity at a rate that more closely matches demand growth and allows the local grid capacity to be built up in modest increments. This provides the owner with operational flexibility and favorably impacts the economics of the plant, as discussed in the previous chapter. Besides the economic advantages of reduced cash outlay and better demand matching, modularization of nuclear plants provides an opportunity to take standardization to a new level. The US did a dismal job at standardizing nuclear plants in our original build-out during the 1960s and 1970s. Nearly every plant was a "one-off" design. Subsequent nuclear build-outs in countries such as France and the Republic of Korea have done a better job at maintaining standardization in their fleet of commercial reactors with an observable gain in improved efficiencies and reduced costs. The new plants being built in the US also anticipate improved standardization, but it is too early to tell if the vendors and new customers will adhere to this goal. Modular plants facilitate the possibility of a highly standardized reactor system, much like the exacting standards of an AA battery, while offering the owner the ability to customize the nonsafety-grade balance-of-plant.

The higher degree of reactor system standardization that SMRs can provide should also accelerate the implementation of fleet-wide system management, similar to what is done in other industries such as jet engines in the aircraft industry. Rolls Royce is able to monitor a large array of operating parameters on thousands of in-service jet engines using sensors and live satellite feeds to predict potential problems before they occur.[11] The ability to monitor the operating condition of each module in real time could further enhance plant safety through the early identification of component aberrant behavior and allow for timely repair or replacement. A further simplification and economic advantage of operating a fleet of standardized modules is the efficiencies gained in maintaining a reduced inventory of replacement parts and qualification of maintenance personnel.

7.3 Siting benefits

In 2009, Senator Lamar Alexander from Tennessee announced an initiative to tackle issues of air quality and climate change by expanding the domestic use of nuclear power.[12] He proposed that the US build 100 new nuclear plants over the next 20 years. Researchers at ORNL set out to determine if there is sufficient qualified land in the

US to site this number of new nuclear power plants—assumed to be large plants. Their efforts resulted in the development of a site-screening tool named the Oak Ridge Siting Analysis for power Generation Expansion (OR-SAGE). The tool combines approximately 30 geographic information system datasets in order to perform multiparameter screening of candidate sites on a 100 m by 100 m grid spanning the entire continental US. One of their first applications of the tool was to determine the amount of land area suitable for SMRs as well as large plants.[13] The most striking result from their study was that there is roughly a two fold increase in the site availability for SMRs relative to large (1600 MWe) plants. The increase results from a number of factors, principally SMR's smaller plant footprint and reduced water usage. Roughly 25% of the total US land area, including at least some amount of land in each of the contiguous 48 states, appears to be suitable for SMRs based on the 10 screening criteria used in the study.

The ORNL researchers observed that water usage was the greatest differentiator for SMRs. An increasing problem in nearly all countries is the availability of sufficient cooling water for effective rejection of the heat from power plants. In the US, roughly 40% of our fresh water withdrawal is for cooling of thermoelectric power stations. Because of thermodynamic efficiencies of the steam Rankine power conversion cycle, a power plant dumps roughly two-thirds of its produced thermal power to the environment, typically an adjacent body of water or the atmosphere via cooling towers. A smaller-sized plant produces less power and therefore rejects less power as residual heat. This means that either a smaller amount of water withdrawal is needed or the cooling water can be discharged from the plant with a smaller temperature increase. This, in turn, allows the SMR to be operated in regions where only small or low-flow rivers are available and also in warmer climates without exceeding water temperature limits. Several SMRs advertise the capability to use air cooling of the condensers rather than water, although the power conversion efficiency decreases by a few percentage points in this case. Even with the reduced efficiency, dry cooling may be the only option in arid regions of the country. The ORNL OR-SAGE analysis indicated that dry cooling of an SMR doubles the amount of suitable land area for SMRs to nearly 60% of the total US land area.

In principle, the smaller building size of the SMR nuclear island may facilitate the use of seismic isolators similar to what is used for conventional buildings in earthquake-prone parts of the world such as Japan. This would substantially eliminate the most significant site-specific design consideration—seismic resilience of the plant—and would further enable greater design standardization. More importantly, the use of seismic isolators would improve the safety of the plant by significantly reducing the probability of seismic-induced damage. The PRISM reactor design developed by General Electric was the first to include seismic isolation of the nuclear island. Other designs have considered this approach, but no nuclear plant has been built with this feature. For multimodule plants, seismic isolation of the individual modules may be a better option than isolation of the entire reactor building. Economic and regulatory factors will determine the ultimate solution.

A significant siting flexibility of SMRs, if it can be realized, is the opportunity to simplify the plant's emergency management plan, including a potential reduction in the plant's emergency planning zone (EPZ). The US Nuclear Regulatory Commission

(NRC) currently sets the EPZ for existing and new nuclear plants at 10 miles for plume exposure and 50 miles for ingestion pathways. However, they allow a license applicant to request a reduced EPZ and have licensed several small reactors with an EPZ of 5 miles. These include the Fort St. Vrain (842 MWt), Big Rock Point (240 MWt), and La Crosse (165 MWt) plants, although none of these plants are still in operation.[14] As discussed in Chapter 5, new SMR designs are expected to have significantly enhanced safety and resilience to accidents, which results from innovative engineering and the extensive use of passive safety systems. They will also have reduced accident source terms by virtue of a smaller reactor core size and may include additional barriers for fission product releases if fuel damage does occur. Hence, it should be possible for an SMR to justify a reduced EPZ size while not increasing risk to the general public.

In recognition of this opportunity and its value, the SMR community has been working through the Nuclear Energy Institute (NEI) to develop and pursue a methodology for deriving the appropriately "optimized" emergency preparedness of an SMR that is commensurate with the plant's relative risk. The NEI submitted a white paper on the topic to the NRC in December 2013, describing a methodology for establishing a "technology-neutral, dose-based, consequence-oriented emergency preparedness framework for small modular reactor sites."[15] The proposed methodology, which was accepted by the NRC in 2015, requires that a license applicant provides a substantiated technical basis for the proposed emergency planning actions, including the extent of the EPZ.

Optimization of the EPZ for SMRs is valuable for a number of reasons. First, an analysis at the Idaho National Laboratory determined that the average cost of establishing a 10-mile EPZ averages about $10 million per plant and adds greater than $2 million to the annual operating cost.[14] A substantial portion of these costs would be saved if a reduced EPZ can be justified. More importantly, a reduced EPZ facilitates siting of the plant closer to population centers without the costly and socially alarming aspects of emergency planning requirements. This is especially important if SMRs are to be successful as an option for distributed power generation for smaller, more remote communities. A reduced EPZ will also help to facilitate broader missions for SMRs, especially process heat missions. In this case, colocation of the reactor with the industrial user is very important to minimize heat losses due to lengthy steam distribution lines. This brings us to the fourth major flexibility that SMRs have to offer: suitability for nonelectrical applications.

7.4 Adaptability to heat applications

A huge attractiveness of small reactors is their flexibility to enter traditionally non-nuclear energy markets. Several energy-intensive applications could significantly benefit from using nuclear-generated heat to replace the combustion of fossil fuels to produce heat. This would not only reduce the emission of greenhouse gases but would also allow the fossil resources to be used as feedstock for producing higher value products such as petrochemicals and plastics. In the US, electricity generation contributes only 40% to our total carbon emissions, so a concerted effort to reduce our total emissions will necessarily require replacing fossil fuels with nonemitting fuels in the industrial sector as well as the electricity generation sector.

The suitability of SMRs for this mission is largely a result of their smallness and modularity and to a lesser extent their siting flexibilities. Primary applications for nuclear heat include the following:

- districting heating,
- water desalination and purification,
- advanced oil recovery processes and oil refining,
- hydrogen production for the enrichment of liquid fuels and eventually fuel cell applications,
- advanced energy conversion processes such as coal-to-liquids and petrochemical production, and
- general process heat for chemical or manufacturing processes.

In all cases, smaller-sized, more robust reactors are better suited for integration with these applications than large plants because of the many considerations discussed earlier. I will defer a review of these reasons until the next chapter, which takes a customer-centric view of SMRs. In this section, I provide examples of the first three nonelectrical applications listed above and focus on the technical aspects of SMRs that facilitate coupling to industrial applications. In some cases, current SMR designs may not be adequate, at least not without significant redesign, and I highlight these as well.

The most logical integration of a nuclear plant with nonelectrical applications is to colocate the two facilities and dedicate the nuclear plant to the industrial facility. The energy coupling may be heat only or a combination of heat and electricity. The latter arrangement is typically referred to as either "cogeneration" or as "combined heat and power" (CHP). The aspect of colocation is what gives SMRs a distinct advantage over large power plants because their smaller output is typically a better match for a single industrial plant. The International Atomic Energy Agency (IAEA) has conducted a number of studies of nonelectrical applications for nuclear plants, including for traditional water-cooled reactors and advanced nonwater-cooled reactors. An especially thorough review on the subject for water-cooled reactors is reported in *Advanced Applications of Water-Cooled Nuclear Power Plants*, issued in 2007.[16]

7.4.1 District heating

Nonelectrical applications are not entirely new to nuclear power. The most common example is district heating. Currently 59 nuclear plants at 21 sites in 9 countries provide district heat to residential and/or industrial customers, including Bulgaria, Czech Republic, Hungary, India, Romania, Russia, Slovakia, Switzerland, and Ukraine.[16] District heating requires relatively low temperature heat, typically 80–150°C. Although steam can be used to transport the heat to the end user, more often water is used because of its higher heat capacity, which reduces heat losses during transport (up to several kilometers in length). Coupling of the reactor to the district heating system is quite straightforward: circulate the output steam from the reactor through an external heat exchanger to transfer the heat to a circulating loop of water. Most often, the heat-generating plant operates in cogeneration mode, that is, it produces electricity as well as the heat supply. In this case, steam is typically extracted from an intermediate stage of the turbine and diverted to the external heat exchanger while the remainder of the steam is used to run the turbine/generator equipment.

While large power plants—fossil and nuclear—can and do support district heating applications, SMRs offer a flexible and scalable option due to their smallness and modularity. Obviously, heat demand is directly related to population size and climate. It also can have a strong seasonal variation. The peak heat load for a small town can be on the order of 10–50 MWth and 800–1200 MWth for a medium-sized city. On the higher end, the peak district heating load in Warsaw is approximately 4000 MWt. This amount of heat could be accommodated by a dedicated large power plant; however, heat transmission lines would become prohibitively long. Multiple smaller heat sources are preferred even in this high-demand market. Also, the modularity of some SMR plants can allow individual modules to be dedicated to either heat production for the districting heating system or electricity production. This feature would improve the simplicity of the balance-of-plant systems and improve the overall plant efficiency.

7.4.2   Water desalination

In terms of ease of coupling with a nuclear plant, water desalination is similar to district heating. However, far fewer commercial reactors, less than 15, have been used to desalinate water. Given that there are roughly 16,000 desalination plants worldwide, nuclear-supplied heat contributes to less than 0.1% of the global desalination capacity.[14] As examples of nuclear desalination, Kazakhstan operated a distillation-type desalination plant at the Aktau site for over 27 years until the reactor was shut down in 1999. India coupled a desalination plant to the Madras Atomic Power Station in 2002 and at the Kudankulam site in 2009. Japan has accumulated the greatest amount of experience with nuclear-driven desalination plants, having operated 10 desalination units at 4 nuclear plant sites before the country-wide shutdown of their nuclear plants in 2011.[17] In contrast to this limited amount of commercial nuclear desalination, nuclear-powered naval vessels routinely use nuclear energy to desalinate sea water.

The choice of desalination method, and hence the options for integrating an SMR to a desalination plant, is determined primarily by the characteristics of the source water and the desired product water. For example, reverse osmosis (RO) technology has the best purification efficiency for many users but is less effective with feedwater that contains a high level of organic materials or that has a high salinity level. Also, the RO product water may require additional treatment to achieve a high target purity. The two common thermal distillation processes, multistage flash distillation (MSF) and multieffect distillation (MED), are much more tolerant of "dirty" or "salty" feedwater and produce high purity water but at a lower efficiency than RO. For this reason, hybrid plants that use a combination of RO and either MSF or MED technologies are becoming more common.

Several SMR vendors advertise their applicability to water desalination. Two designs have already received design approval from their regulators: the SMART reactor developed in the Republic of Korea and the barge-mounted KLT-40S reactor developed in the Russian Federation. Both are being marketed as cogeneration plants for producing electricity and water. Similar to cogeneration of electricity and heat for district heating, the heat for the desalination plant is typically extracted from the low-pressure turbine stage at a temperature of 100–125°C and circulated through an external heat exchanger.

In a study at NuScale Power, we teamed with Aquatech International to investigate technical options for coupling a NuScale SMR plant to RO, MSF, and MED desalination plants.[9] We looked at three options for providing heat to the desalination plant: high pressure (HP) steam taken before insertion into the turbine, medium pressure (MP) steam taken from the intermediate turbine stage, and low pressure (LP) steam taken from the exhaust end of the turbine. Each option produced different water purification efficiencies in the different desalination technologies. In the case of the RO plant, waste heat from the nuclear plant was used to preheat the RO feedwater to further improve its efficiency.

Figure 7.2 shows the relationship between the two products of the combined nuclear and desalination plant: electricity and water. The figure shows the clear advantage of the RO process in terms of water produced due to its high conversion efficiency. This comes at the expense of water quality since the RO process yields lower water purity than the thermal distillation processes. For the thermal desalination processes, the plant's electrical output is higher when lower pressure steam is used. The trade-off is a successive reduction in operational flexibility as the motive source is changed from main (HP) to extraction (MP) to exhaust (LP) steam.

The modularity of the SMR helps to restore some flexibility. In this case, a single SMR module of 50 MWe/160 MWth size is sufficient to generate 200,000 m³/day of clean water from an RO plant—a reasonable size for a commercial desalination

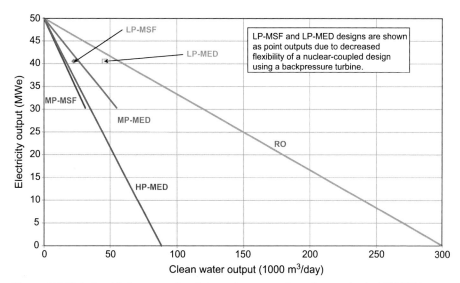

Figure 7.2 Relationship between electricity and water products from a single 160 MWth SMR module coupled to a variety of desalination processes.[9]

plant—whereas four modules would produce a similar amount of water using an MED plant and six modules using an MSF plant. In the latter two cases, each module is also producing sellable electricity, so local prices for electricity and water will influence the most favorable balance of electricity versus water production. The point is that the modularity of the SMR gives the plant owner the flexibility to evaluate this balance and adjust the balance as market conditions change.

7.4.3   Oil recovery and refining

The third example I offer regarding the flexibilities afforded by SMRs to nonelectrical applications is for the recovery and refining of oil. Most of the easily accessible crude oil in the US has already been depleted, and oil companies are utilizing energy-intensive processes to increase oil recovery from existing fields, extraction from new formations such as tar sands, or extraction from nontraditional sources such as oil shale. In the case of enhanced recovery from tar sands, 90% of the energy usage is steam, which is used in a process called "steam-assisted gravity drain." The steam is injected in situ to reduce the viscosity of the oil, which can then be pumped out using conventional methods. The steam quality is generally quite low and dirty. In the case of oil shale, the oil is actually contained in the sedimentary rock as kerogen, which is converted to light oil and other products by slow heating.[18] The heating can be either in situ as part of underground heating operations or ex situ after it has been mined and trucked to a central facility for heating and oil extraction.

Temperature requirements for enhanced oil recovery processes generally range from 250 to 350°C. Cogeneration is generally preferred due to the need for a modest amount of electricity for pumping operations and general housekeeping functions. Long disruptions in the heat-up process would become expensive if the rock formations are allowed to cool down significantly. So in principle, a small, modular heat source such as an SMR should be a good technical solution for oil recovery. However, challenges appear to be dominated by economic and logistical considerations. First, the oil recovery application requires a geographically distributed array of very small energy sources. This might dictate that the nuclear modules be deployed as single or few-module clusters, which would dramatically increase the cost of construction, operations, and security.[18]

The other logistical disconnect for in situ oil recovery is the longevity of the field operations in relation to the anticipated SMR plant lifetime. Using enhanced recovery processes to extract heavy oil or tar sands oil is likely to fully deplete a specific field in 10–15 years. This depletion time can be increased with the use of horizontal drilling, which extends the reach of a well. Even so, the demand for a heat source at a fixed location is likely to be significantly shorter than the typical 60-year lifetime of a nuclear plant. Currently, oil companies use mobile natural gas units to provide the energy. Comparable nuclear options might include the development of a mobile nuclear plant or a less enduring plant with a 10- to 20-year design lifetime. These options have their own set of challenges requiring lengthy research and development efforts. In the case of oil shale, there are indications that the deposits are sufficiently massive and the heating process sufficiently protracted that harvesting the oil from these formations may require many tens of years and hence be a better match for a nuclear plant with a traditional design life.

Although these economic and logistical considerations cause the nuclear option for in situ oil recovery to be quite challenging, ex situ recovery, that is, the shale oil is mined and processed elsewhere, is a promising application that overcomes the distributed and migratory issues of in situ recovery. Furthermore, oil recovered from tar sands is of sufficiently low quality that it requires processing in upgrader facilities located near the oil fields. Upgraders are basically in-field refineries that can service a large oil recovery area. As the local recovery operations migrate to new areas of the larger field, the oil is piped over progressively longer runs to the upgrader. The upgrader has a much longer lifetime and energy demand characteristics similar to finishing refineries, which appear to be an even more promising fit for SMR application.

The energy requirements of a refinery represent a more practical and potential application of an SMR. Refineries are large, energy-intensive industrial complexes with extended lifetimes similar to nuclear power plants. Although the initial design lifetime of refineries may be 20 years, they are frequently upgraded as technology improves or product markets evolve and typically operate for several decades. One of the longest running refineries in the US is the Casper Refinery near Rawlins, Wyoming, which has been operating for 90 years. Also, many refineries are in less populous areas and have industrial exclusion zones. In 2007, there were 145 US refineries with the average refinery using roughly 650 MWth. Some of the largest refineries can use in excess of 2000 MWth.[19]

An attractive feature of SMRs for this application, at least multimodule SMR plants, is the potential for staggered refueling of modules. Many refinery processes become very inefficient if disrupted and therefore have a high reliability requirement. A multimodule SMR plant provides for redundancy and the continuous availability of heat supply. However, actual coupling of heat from the SMR to the various process flows in the refinery is much less straightforward than for district heating and desalination. Individual processes require different temperatures and different rates of heat delivery. Some processes use direct fired heat, which is difficult to replicate with an external heat source such as a nuclear reactor. Also, a refinery waste product called refinery fuel gas is produced and can be used to provide some of the fired heat needed by the plant without incurring additional fuel costs.

In collaboration with Fluor Corporation, NuScale Power investigated the extent of potential coupling between a nuclear plant and a typical large-size oil refinery.[8] As a case study, we selected a refinery capable of processing 250,000 barrels/day of crude oil to produce diesel fuel, gasoline, petroleum coke, and other petroleum products. Anticipated energy demands for this scale of refinery are listed in Table 7.2. Six NuScale modules are sufficient to provide the required 250 MWe of electricity for the refinery as well as the house load for the NuScale plant. To determine how many modules are needed to meet the nonelectrical energy demands required a review of the detailed process flow characteristics of the refinery. No credit could be taken for heat provided by the refinery fuel gas since this is a by-product of refinery processes and is basically free. Also, the use of natural gas in a methane reforming process appeared to be the most efficient process for hydrogen production. The

Table 7.2 **Primary energy demands for typical refinery producing 250,000 barrels per day**

Traditional heat source	Energy demand (MBtu/h)	Replaceable by SMR module (MBtu/h)
Natural gas		
• For 250 MW of electricity	1900	1900
• For H_2 production	4100	No
• For fired heaters	1800	1660
• For pilot lights	140	No
Refinery fuel gas		
• For fired heaters	2000	No

study concluded that approximately 1660 MBtu/hr of the 1800 MBtu/hr heat load in the refinery that is normally supplied by combusting natural gas could realistically be met with nuclear-supplied steam. This would require four NuScale modules in addition to the six modules needed to supply the electrical load. Hence, with these assumptions, a 10-module NuScale plant is the appropriate match to a 250,000 barrels/day oil refinery and results in a 36% (190 MT/hr) reduction in CO_2 emissions from the refinery.

7.4.4 Hybrid energy system applications

I was introduced to the topic of hybrid energy system (HES) about 5 years ago by a small, enthusiastic band of chemical engineers at Idaho National Laboratory (INL) led by Steve Aumeier and later by Richard Boardman. I was slow to embrace the concept, perhaps because chemical engineers speak a different language than nuclear engineers, and I lacked an interpreter. I now believe that HESs will be a part of our evolving energy future. In principle, an HES applies intrinsically to all energy sources, large and small. I include the topic in this chapter because there are numerous reasons why small generating units allow greater flexibility in the realization of HES. A good overview of HESs and especially the benefits of SMRs for this application is given by Shannon Bragg-Sitton of INL in Chapter 13 of the *Handbook of Small Modular Nuclear Reactors*[20] and in an INL technical report.[21]

An HES is a collection of multiple energy sources and multiple energy uses integrated into an optimized system that enables each source, production process, and storage interface to operate in its "sweet spot," that is, in a manner that maximizes its efficiency and economics. It brings together all of the cogeneration applications discussed above and much more. Figure 7.3 captures my simplified representation of an HES. In this example, four energy generators (two that produce electricity directly and two that produce heat) are coupled to two energy users: the electrical grid and a desalination plant. The important aspect of this diagram is that the energy sources are tightly coupled, that is, they combine their energy outputs prior to distribution on the grid or the process heat plant. It is

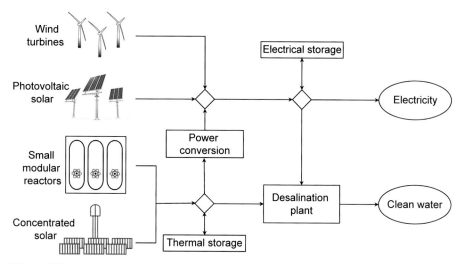

Figure 7.3 Notional concept of a hybrid energy system showing four different energy producers and two energy-derived products.

this coupling "behind the curtain" that allows for true optimization of the various energy outputs.[22] Collectively, the components make up a highly dynamic system with variable generators and variable loads.

If properly implemented, an HES should allow for each functional component to operate in its optimum way. While SMRs can accommodate a load-following operation, it is economically and operationally better to operate them continuously at full power. In an HES, variations in the grid load and intermittent generators can be accommodated by switching electricity and heat generation to variable product processes. From the perspective of the SMR, load-following is replaced by load-switching or what Bragg-Sitton calls "load dynamic" operation. This may seem like passing the buck from the generators to the users, but there are many industrial processes that can be easily cycled or operated at variable production rates. There are, however, many other processes that are highly sensitive to heat delivery rates. These processes can still be incorporated into an HES if done so in a way that ensures continuous energy input. Giorgio Locatelli discusses a good example of this distinction in his paper on load-following verses cogeneration, in which he compares dynamic energy delivery impacts on biofuel production and water desalination.[23]

Modularity is the key to dealing with the challenges of this complex optimization problem. Wind and photovoltaic solar are already highly modularized, and concentrated solar will tend to be modular due to light collection efficiencies. SMRs allow the nuclear component to be modular as well. The same principle is true on the user side of the system; fortunately, many industrial processes are already modularized. Other benefits of SMRs for HES cited by Bragg-Sitton include small unit size, incremental build-out, and siting flexibilities.

This concludes Part Two—the portion of the book that is largely devoted to explaining in modest detail how SMRs are different from large plants and the features that enable them to achieve higher levels of safety, enhanced affordability, and expanded flexibility for diverse energy applications. In the third and final part, I will address these benefits from a customer's perspective and provide my assessment of the several challenges that remain for the eventual deployment of SMRs as well as the opportunities that these challenges could catalyze. In the final chapter, I will revisit the opening question of SMRs: nuclear power—fad or future?

References

1. Carelli MD, Ingersoll DT. *Handbook of small modular nuclear reactors*. Cambridge (UK): Woodhead Publishing; 2014.
2. *Advances in small modular reactor technology development*. International Atomic Energy Agency; 2014. Available at: http://aris.iaea.org.
3. *Global database of operational generation plants*. 3rd ed. Research and Markets, Ltd; 2006.
4. *Technical and economic aspects of load following with nuclear power plants*. Nuclear Energy Agency; June 2011.
5. Ingersoll DT, et al. Can nuclear power and renewables be friends? In: *Proceedings of the international conference on advances in power plants, Nice, France, May 3–6, 2015*.
6. Greene SR, Flanagan GF, Borole AP. *Integration of biorefineries and nuclear cogeneration power plants – a preliminary analysis*. Oak Ridge National Laboratory; 2008. ORNL/TM-2008/102.
7. Li N. A paradigm shift needed for nuclear reactors: from economies of Unit scale to economies of production scale, In: *Proceedings of the international congress on advanced power plants, Tokyo, Japan, May 10–14, 2009*.
8. Ingersoll DT, Colbert C, Bromm R, Houghton Z. NuScale energy Supply for oil recovery and refining applications, In: *Proceedings of the 2014 international congress on advances in nuclear power plants, Charlotte, NC, USA, April 6–9, 2014*.
9. Ingersoll DT, Houghton ZJ, Bromm R, Desportes C. NuScale small modular reactor for co-generation of electricity and water. *Desalination* 2014;**340**:84–93.
10. Carlsson J, et al. Economic viability of small nuclear reactors in future European cogeneration markets. *Energy Policy* 2012;**43**:396–406.
11. *Rolls Royce engine health management system*. 2015. Available at: http://www.rolls-royce.com/about/technology/enabling_technologies/engine-health-management/.
12. Sen. Alexander L. (R, TN). Build 100 new nuclear power plants in 20 years for a rebirth of industrial America while we figure out renewable electricity, address to the Tennessee Valley Corridor Summit, Oak Ridge, TN, May 27, 2009.
13. Belles RJ, Mays GT, Omitaomu OA, Poore WP. *Updated application of spatial data modeling and geographical information systems for identification of potential siting options for small modular reactors*. Oak Ridge National Laboratory; 2012. ORNL/TM-2012/403.
14. *Opportunities in SMR emergency planning*. Idaho National Laboratory; 2014. INL/EXT-14-33137.
15. *Proposed methodology and criteria for establishing the technical basis for small modular reactor planning zone*. Nuclear Energy Institute; Submitted to the US Nuclear Regulatory Commission on December 23, 2013.
16. *Advanced applications of water-cooled nuclear power plants*. International Atomic Energy Agency; 2007. IAEA-TECDOC-1584.

17. *Nuclear desalination*. World Nuclear Association; July 2013. Available at: www.world-nuclear.org/info/Non-Power-Nuclear-Applications/Industry/Nuclear-Desalination.
18. Curtis D, Forsberg CW. Light-water-reactor arrays for production of shale oil and variable electricity. *ANS Trans* 2013;**108**.
19. Konefal J, Rackiewicz D. *Survey of HTGR process energy applications*. May 2008. MPR-3181.
20. Bragg-Sitton S. Hybrid energy systems using small modular reactors. Chapter 13. In: *Handbook of small modular nuclear reactors*. Cambridge (UK): Woodhead Publishing; 2014.
21. Bragg-Sitton S, et al. *Value proposition for load-following small modular reactor hybrid energy systems*. Idaho National Laboratory; 2013. INL/EXT-13-29298.
22. Ruth MF, et al. Nuclear-Renewable hybrid energy systems: opportunities, interconnections, and needs. *Energy Convers Manage* 2014;**78**:684–94.
23. Locatelli G, Boarin S, Pellegrino F, Ricotti M. Load following with Small Modular Reactors (SMR): a real options analysis. *Energy* 2014. Available at: http://dx.doi.org/10.1016/j.energy.2014.11.040.

Part Three

Promise to reality

Customer buzz about small modular reactors

The preceding three chapters approach small modular nuclear reactors (SMRs) from a technologist's point of view and may leave the impression that SMRs are merely a classic "technology push," a label sometimes assigned by critics to substantiate their prediction of SMRs as a short-lived fad. In reality, there has been and continues to be significant interest in SMRs by potential customers worldwide. So in this chapter, I approach the case for SMRs from the customers' perspective based largely on my direct engagement with interested customers over the past several years. I also provide a cross-link between the various attributes of SMRs discussed previously and the specific energy needs, expectations, and constraints of several different classes of potential customers. Since the leading intent of SMRs is to extend the nuclear energy option to new types of customers, many of these customers have very different requirements and expectations than the traditional large-grid, base-load customer. I will not belabor the specific SMR benefits since these are covered thoroughly in previous chapters. However, a customer-centric presentation of SMR benefits should help explain why there has been so much interest in SMRs from so many different constituencies and why SMRs are not merely a "technology push." In short, this chapter should help to answer the questions "Why SMRs?" and "Why now?" from the viewpoint of a buyer. I start with developing countries because I believe that nuclear power, as facilitated by SMRs, has the greatest potential to improve their future.

8.1 Emerging countries

Projections of future energy demands are about as reliable as typical economic projections—useful but almost certain to be wrong. So while I do not take seriously the precise numbers from these projections, some clear trends related to future energy demands are indisputable:

- the world population will continue to grow,
- countries with low qualities of life will seek to improve their economies through industrialization,
- developed countries will seek to maintain or further improve their quality of life, and
- increasing water scarcity will require nearly all countries to use more energy to produce clean water.

These combined factors will yield a rapid and sustained increase in the demand for energy worldwide. A projected doubling of global energy demand by 2040–2050 is common among analysts and most agree that a high percentage of this (75–80%) will come from countries with emerging economies. Overlaid on this picture is a rapidly

Small Modular Reactors. http://dx.doi.org/10.1016/B978-0-08-100252-0.00008-2

permeating concern for global climate change impacts resulting from the widespread emission of greenhouse gases (GHG) such as carbon dioxide and nitrous oxide, as well as general air quality concerns from the emission of acidic gases and toxic metals such as sulfur dioxide and mercury. The combination of the rapid growth of energy consumption, decreasing energy resources, and increasing constraints on energy quality creates a serious dilemma for developing countries. Every industrialized country today went through a period of poor air quality and environmental abuses in order to produce the abundant energy needed to feed their industrialization. China is passing through this apparent "rite of passage" now and is just beginning to deal with air quality issues. The challenge to countries that have not yet made the jump to industrialization is that international pressures and formal protocols may no longer tolerate wholesale use of dirty fuels. So how will these emerging countries secure the energy they need to achieve their economic and social goals?

In 2006 when the US Department of Energy (DOE) introduced their Global Nuclear Energy Partnership (GNEP) program, I jumped at the opportunity to engage with a specific component of GNEP that focused on the international deployment of small power reactors, which became known as the Grid-Appropriate Reactors program. I had already fully embraced the merits of SMRs through my participation with the International Reactor Innovative and Secure (IRIS) project and gained considerable insight into international markets due to the multinational nature of the IRIS Consortium. I was convinced that the first SMR purchase would require a highly motivated customer, one that had few or no other choices, and that this would be a small developing country. Indeed, the case for SMRs in several countries with emerging economies is quite compelling. However, after visiting several of these countries and participating in numerous multinational meetings hosted by the International Atomic Energy Agency (IAEA), I came to appreciate the formidable challenges associated with introducing nuclear power into these countries and some surprising expectations associated with SMRs specifically.

Many developing countries are experiencing rapid electricity demand growth (8–10% per year) but have exhausted their indigenous energy resources. Given the choice of importing fossil fuels such as oil, coal, or natural gas or initiating nuclear power to meet their demand growth, many countries are choosing to pursue nuclear power. A representative from Ghana was the first to approach the DOE for assistance in deploying SMRs in Ghana as part of the GNEP program. Their only indigenous energy resource is hydroelectric power, but a sustained drought had seriously lowered the water level in their many reservoirs. They were faced with the inevitable choice of which new energy resource to import. Ghana's initial study of energy options concluded that nuclear power was a viable candidate; however, there are only a few locations on their coast that can accommodate a large nuclear plant.[1] SMRs appeared to be a much better match for their geographic distribution of communities throughout the country and their limited infrastructure. Also, Ghana's total generating capacity at that time was less than 2000 MW,[2] which precluded a large plant option based on grid stability and spinning reserve considerations.

Ghana's energy situation is representative of several countries that eventually approached the DOE for help in pursuing SMRs. The majority of countries seeking to initiate nuclear power programs are severely constrained on the size of power plants that they can afford and operate on their local grids. In recognition of this, the IAEA conducted a 2-year program entitled "Common User Considerations" to explore the needs and constraints of developing countries. The study concluded that roughly 60% of the countries are limited to considering plant sizes less than 600 MWe when accounting for local grid capacity and financial constraints.[3] Collectively, these countries represent more than half of the anticipated global energy demand growth in the next few decades and therefore represent a significant customer base for new SMR designs.

It was my participation in the IAEA's Common User Considerations study that gave me the best collective view of the key expectations of the developing countries that were seeking to initiate a nuclear power program. The first surprise was their insistence on considering only proven (demonstrated) technology, which to them meant commercial plant designs that had already been built and operated for at least a few years somewhere, preferably in the vendor's country. This presented an immediate dilemma for them since no contemporary SMR design had been built anywhere, and at the time of the study, none had even been licensed by a regulator. Related to this was their desire to not be treated as "guinea pigs," that is, they expected evidence that new plant designs were of value to and deployed in the vendor's country before being exported. Although this is a valid expectation, there have been some notable exceptions to this domestic-first goal. Specifically, construction of the first Areva EPR plant was in Finland, and the first two dual-unit Westinghouse AP-1000 plants were constructed in China. In both of these cases, however, subsequent plant constructions were initiated in the vendors' countries.

A third major requirement of the developing countries was that a supply of fuel, components, and services be assured for the lifetime of the plant. This may seem like an odd expectation, but many smaller countries recognize their vulnerability to political leveraging by larger countries. While legal agreements are a reasonable start, they prefer commercial arrangements that provide multiple supplier options as a more reliable hedge against the disruption of parts and services. The final key expectation was that the reactor design be licensed by a reputable regulator, presumably the regulator in the vendor's country. This requirement reflects the limited regulatory infrastructure of the receiving country and is also consistent with their domestic-first expectation. It is still the purchasing country's responsibility to license and regulate operating reactors within their boundaries, but purchasing a previously certified design adds an additional level of confidence.

The common denominator of the many requirements and expectations articulated by the developing countries was that they need their first nuclear plant project risk to be extremely low—essentially zero. This is understandable when you consider that the pursuit of a nuclear plant by many of these countries is a "bet the country" endeavor in terms of both financial limitations and national developmental goals. This expectation may require a more transparent and intimate engagement of the vendor with the customer to provide the additional confidence that the new plant will not only operate reliably but also that long-term support is assured.

The actual size of the potential international market for SMRs, including specific countries, is difficult to assess. The IAEA quotes numbers ranging from 35 to 45 for the number of countries interested in pursuing a nuclear power program. However, this number is based on self-identification by those countries and does not necessarily reflect near-term intent or viability. Geoffrey Black and a group of researchers at the Energy Policy Institute (EPI) conducted a study using a systematic and quantitative approach to assess the potential for global SMR deployment.[4] Conducted in cooperation with the IAEA, the study was nondiscriminatory, that is, it evaluated all countries for which there was sufficient reputable data. The initial list of 214 countries was reduced to 97 based on 4 screening criteria that include gross domestic product, per capita income, size of electrical grid, and being a signatory to the Treaty on the Nonproliferation of Nuclear Weapons. The remaining 97 countries were scored and ranked based on 15 quantitative evaluation criteria, which included factors such as financial and economic status, technology infrastructure, and government and regulatory infrastructure. Table 8.1 provides a list of the highest-ranked 30 countries.

There are several surprises in the ranking results from the EPI study. Interpreting the results requires an understanding of the 4 screening criteria and the 15 evaluation criteria, for which I defer you to the referenced paper. One result that I do not find surprising is that Singapore ranked highest. I have listened to or read material on the energy situation in several countries. I was especially struck by a presentation on Singapore's circumstances and noted that Singapore represents a classic case for SMRs. The country is a modestly sized, densely populated island. It has a peak electricity demand of roughly 7 GWe and a single grid connection (to Malaysia). There is no wind and very little space for solar panels; consequently, they import all of their energy. In terms of the nuclear power option, evacuation is not an option, so the nuclear plant must be nearly bullet-proof in terms of safety and robustness. Because of the isolation factor, reliability is also critical, which implies a modular plant with redundant units.

While Singapore's case is extreme, it typifies many countries in terms of their requirements for nuclear power, which include high levels of safety and resilience

Table 8.1 Countries that ranked the highest in the EPI study for the suitability of deploying SMRs[4]

Rank	Country	Rank	Country	Rank	Country
1.	Singapore	11.	Saudi Arabia	21.	Finland
2.	Qatar	12.	Israel	22.	Chile
3.	Luxembourg	13.	Germany	23.	Slovenia
4.	Ireland	14.	Belgium	24.	Panama
5.	Republic of Korea	15.	Austria	25.	US
6.	Netherlands	16.	Estonia	26.	United Kingdom
7.	United Arab Emirates	17.	Trinidad and Tobago	27.	Denmark
8.	Oman	18.	Thailand	28.	Sweden
9.	Bahrain	19.	Switzerland	29.	Australia
10.	Malaysia	20.	Cyprus	30.	Czech Republic

appropriately sized for their demand and grid infrastructure as well as modularity for enhanced reliability. For many of the countries, the greater siting flexibilities and adaptability to process heat applications also have been cited as favorable attributes of SMRs. One potential benefit of SMRs for these countries is that smaller component sizes and simplified systems of some SMRs may allow for faster and more extensive localization of the country's labor force and manufacturing capabilities.[5] While I agree with this assertion, I label it as a "potential" benefit because it runs contrary to the assumption by many SMR vendors that anticipate SMRs being manufactured in their facilities and exported to other countries. Several political and economic factors will ultimately dictate the optimum balance of the supply chain between the vendor's country versus the buyer's country, including factors such as material and labor costs, labor skills, and regulatory oversight.

The desirability of enabling the expansion of nuclear power to more countries has been openly challenged by some, particularly those concerned with nuclear proliferation. Indeed, the global expansion of nuclear energy, enabled in part by SMRs, comes with significant responsibilities for both the technology providers and users. As more reactors are built and more countries join the nuclear energy community, it is vital that the highest levels of safety and security be implemented for new plants and that concerns regarding nuclear weapon proliferation be sufficiently addressed. Countries that already have a substantial investment in nuclear power must work together to ensure that the growing nuclear community respects the need to protect those investments through safe and secure operations of all nuclear plants. We should facilitate and assist these countries along their development path by providing a viable option for clean, abundant, affordable power using the same technology that has helped us to achieve our high quality of life. It would be irresponsible and hypocritical for us to do otherwise.

The IAEA is centrally positioned to facilitate the expanded use of nuclear power worldwide. In addition to being chartered to ensure international safeguarding of nuclear materials, they have numerous assistance programs, topical guidelines, evaluation tools, and dialogue forums to help existing and embarking nuclear countries responsibly initiate and maintain their nuclear programs. About the same time as the Common User Considerations study, the IAEA issued a guide for newcomer countries entitled *Milestones in the Development of a National Infrastructure for Nuclear Power*.[6] This document organizes the initiation of a new nuclear program into three major milestones: (1) making a knowledgeable commitment to a nuclear power program, (2) inviting bids for the first nuclear power plant, and (3) commissioning and operating the first nuclear power plant. The activities needed to achieve these three milestones are further subdivided into 19 specific infrastructure elements, which are listed in Table 8.2. I have also included in Table 8.2 my assessment of the impact that SMRs may have on each of the infrastructure elements in terms of facilitating the introduction of nuclear power into embarking countries. Six of the 19 infrastructure elements are not significantly impacted by the choice of SMRs relative to large nuclear plants; however, many of the remaining elements may be improved through the deployment of SMRs. The actual impact will depend substantially on the specific SMR design selected, which can vary widely in features and technologies.

Table 8.2 **The 19 infrastructure elements of the IAEA Milestones report and impact of deploying SMRs**

Infrastructure Element	Potential SMR Impact
National Position	No impact
Nuclear Safety	Enhanced levels of safety and greater accident resilience should facilitate faster acceptance by stakeholders.
Management	Standardization of nuclear modules may result in improved cross-sharing of management experience and greater management efficiency.
Funding and Financing	Reduced capital cost (<$3B) is easier to finance. Phased build-out of plant can further reduce maximum debt.
Legislative Framework	No impact
Safeguards	Some SMRs may require non-traditional approaches to implementing safeguards.
Regulatory Framework	Licensing reviews may be accelerated or delayed by specific SMR technologies and features.
Radiation Protection	No impact
Electrical Grid	Can be deployed on smaller grids and require less reserve capacity. Also may be less sensitive to reliability and availability of off-site power.
Human Resource Development	Peak construction workforce and normal operations workforce significantly reduced. May also avoid large transient workforce for refueling operations.
Stakeholder Involvement	No impact
Site and Supporting Facilities	Can expand the number of acceptable sites due to smaller footprint, lower water usage, and lower transmission requirements.
Environmental Protection	Allows for geographically distributed power production but may require additional environmental assessments.
Emergency Planning	Can result in simplified emergency planning and reduced evacuation zone.
Nuclear Security	Intrinsic design features may provide additional barriers for security and limit vulnerabilities for sabotage.
Nuclear Fuel Cycle	No impact
Radioactive Waste Management	No impact
Industrial Involvement	Simplified designs reduce number of safety-grade components and will allow more diverse supply chain including increased localization.
Procurement	Smaller components and higher level of standardization can simplify supply chain.

8.2 Domestic utilities

With the embarking countries appearing to be quite interested in SMRs but having very protracted schedules for initiating nuclear power in their countries, the DOE turned its attention to assessing the interest of US utilities for SMRs. I continued to support the DOE as it transitioned from the Grid-Appropriate Reactors program to the domestically focused Small Modular Reactor program during 2009–2011. The emerging countries made it clear: they expected our domestic use of SMRs, or at least licensing of US-developed designs by the NRC, before they would consider them for deployment in their countries. Licensing by the NRC still implies a domestic customer since the NRC allocates its resources based on domestic priorities. To our surprise, there was significant interest in SMRs by US utilities—small and large, nuclear and non-nuclear.

Fortuitous timing may have contributed to the high level of domestic interest in SMRs that we encountered. The economic crisis was in full swing, and financing options for large projects were becoming quite scarce. Also, concerns regarding global climate change were solidifying with the implication that the continued use of fossil fuels might become unacceptable in the not-so-distant future. I anticipated that we would see the greatest interest coming from the smaller utilities and rural cooperatives since the case for SMRs is as compelling for them as for the small emerging countries we had been pursuing. In fact, the utility structure in the US has the outward appearance of being like many small "countries" with limited demand, resources, and infrastructure in specific service regions. Indeed, there was a significant amount of interest voiced early by the smaller utilities. But like the emerging countries, these utilities had a very low tolerance for project risk and SMRs appeared to be too risky without demonstrated experience. To our amazement, it was a large utility with substantial nuclear experience that courageously stepped forward to endorse and pursue building the first domestic SMR: the Tennessee Valley Authority (TVA).[7]

Despite the fact that TVA already had five large nuclear plants in operation and a sixth plant underway, Watts Bar Unit 2, the agency quickly embraced several key benefits of SMRs. A specific feature that attracted TVA to consider SMRs is the opportunity for small, incremental capacity addition. The economic crisis that hit the country in 2008 noticeably reduced electricity demand in the TVA service region, and it was unclear how quickly the demand might rebound. Adding module-sized units of new capacity would allow them to better match the changing demand profile. Also, TVA was already beginning to anticipate federal policies that would favor clean energy options and potentially force the retirement of coal plants. In a 2012 discussion paper by the Brattle Group,[8] they anticipate that as much as 30 GWe of coal-generated electricity could be retired soon in response to new clean air standards implemented by the US Environmental Protection Agency and that this number could be up to five fold higher, depending on other potential fuel pricing and carbon penalty policies. The plant footprint and electricity output from an SMR is a much closer match to many coal plants, especially the older, less efficient plants that are likely to be closed first. Figure 8.1 shows the size distribution of coal plants in the US and also groups them into plants that have been operating for less than or greater than 55 years.[9] Nationwide,

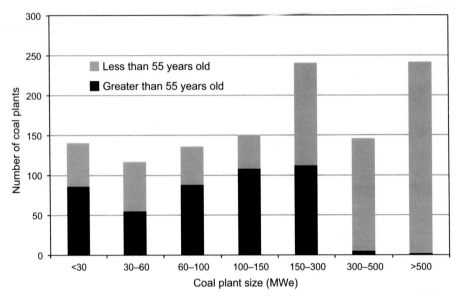

Figure 8.1 Size distribution of US coal plants grouped by their years of operation.[8]

two-thirds of all coal plants have a plant output of less than 300 MWe, and over 90% of the plants that are older than 55 years are below this output level.

TVA quickly identified the Clinch River site in Oak Ridge, Tennessee, as the preferred site for two reasons: (1) it had been environmentally qualified as a nuclear plant site for an earlier advanced reactor project, the Clinch River Breeder Reactor, and (2) the site is adjacent to the Oak Ridge National Laboratory (ORNL), which was highly interested in securing the output from the SMR in order to meet its federally-mandated clean energy goals. The TVA selected the mPower SMR design being developed by Babcock & Wilcox as their reference SMR and began initial site characterization activities. I defer discussion of this particular SMR project until later in the chapter since it crosses over with another type of domestic customer: the US government.

The early embracing of SMRs by TVA appears to reflect the sentiments of many of the large US utilities that operate commercial nuclear reactors. In 2013, the Center for Strategic and International Studies issued a commentary by George Banks entitled, "Why utilities want small modular reactors."[10] Speaking for the broader utility community, Banks cites several key factors, including the need to replace many tens of gigawatts of base-load electricity generation due to coal plant retirements, the improved financing options available with smaller and scalable SMR designs, the advantages of distributed energy generation for stabilizing the grid, their incorporation of post-Fukushima learnings directly into the designs, and their greater flexibility in siting options.

The prospects of SMRs replacing existing coal plants have been cited frequently in the industry press, but some people challenge whether or not this is a viable option. Researchers at ORNL addressed this issue using their newly developed Oak Ridge Siting Analysis for Power Generation Expansion (OR-SAGE) software tool to evaluate the suitability of several coal plant sites for the siting of a small nuclear plant.[11] They used coarse-level screening criteria to produce a manageable number of coal plant sites and then evaluated each site based on several site evaluation criteria such as population density, water availability, and land usage. Their analysis showed that nearly 80% of the coal plant sites scored favorably for SMR siting, which underscores the assumption that coal plants and small nuclear plants share many of the same siting considerations.

In 2014, a public consortium of 44 community-owned utilities servicing eight states in the northwestern US announced its interest in building an SMR. Specifically, the Utah Associated Municipal Power Systems (UAMPS) has teamed with Energy Northwest, a 27-member public consortium servicing the state of Washington, to pursue building a NuScale SMR plant in Idaho. Although the anticipated closure of numerous coal plants was cited as the basis for needing clean replacement power, UAMPS' primary motivation in considering SMRs is their enhanced level of safety.[12] In a related article, Dale Atkinson of Energy Northwest explained their interest in pursuing SMRs:

> *Public utilities in the Northwest and elsewhere are looking for a carbon or fossil hedge. Nuclear generation provides that hedge, and SMR technologies incorporate the lessons learned over several decades of operating similar sized US Navy reactors as well as traditional sized commercial reactors.[13]*

8.3   Process heat users

While interest in SMRs by developing countries and domestic utilities emerged quite spontaneously, interest by nonelectrical customers has been slow to materialize. This may be more of a US issue since the US has virtually no domestic experience with using commercial nuclear power for process heat applications. Another reason may be that existing heat applications tend to be highly optimized for the fuels already at hand: primarily fossil fuels. Some examples were discussed in the previous chapter such as for biomass refineries and oil refineries. In these cases, low-grade refinery waste products provide a cost-effective combustion fuel for generating heat. So it cannot be assumed that just because a process heat application requires heat and a nuclear plant, large or small, produces heat, that the two can be easily integrated. When they can, however, the SMR attributes discussed in previous chapters provide some advantage to these nontraditional applications, specifically in terms of the avoidance of carbon emissions, better heat demand matching, flexibility in siting, and enhanced safety, especially as it impacts the simplification of emergency planning.

A less obvious driver for the reluctance of process heat customers to openly engage in the SMR debate may not be based on technical considerations but rather on business sensitivities. For example, a colleague at ORNL spoke openly at a public conference regarding a study that he had conducted for a leading oil-producing company regarding the potential use of nuclear heat to produce hydrogen, which is an important feedstock in the enriching of crude oil for the production of gasoline. Early the next day, the ORNL Laboratory Director received a call from a senior executive of the oil company. The oil company executive made it clear that his company did not wish it to be publically known that they were having such discussions for reasons of commercial competitiveness and shareholder sensitivities. This is just one piece of anecdotal evidence, but it may reflect a broader reluctance of these nontraditional customers to openly pursue SMRs or nuclear in general.

One exception to the relatively quiet posture that process heat customers are taking regarding SMRs is the state of Wyoming. Wyoming has been partnering with researchers at Idaho National Laboratory (INL) to explore how nuclear-supplied heat can contribute to the repurposing of the state's coal resources.[14] Their studies conclude that the integration of a nuclear plant with an advanced coal conversion plant is technically and economically viable for creating a carbon conversion industry in the state to produce high-value products such as synthetic transportation fuels and carbon-based chemicals. Initial studies focused on the use of small, high-temperature reactors, although more recent studies have also shown the viability of small water-cooled reactors for this application. In addition, the state is exploring with INL the incorporation of SMRs, local wind farms, and coal conversion plants into a potential demonstration of hybrid energy systems. In this case, SMRs provide the needed size matching and siting flexibilities to facilitate the hybrid energy system.[15]

The bottom line for process heat customers is that their interests in SMRs are still largely unknown. In general, the nuclear community has not done a very good job at reaching out to these potential customers. Although I have personally participated in several studies to explore the integration of SMRs with process heat applications, I am guilty, as well as many of my colleagues, in publishing the results in nuclear-related journals and presenting results at nuclear industry conferences. If we are to be successful at educating process heat customers about the options and opportunities for SMRs for their applications, we will need to do a better job at reaching them more directly in *their* established forums.

8.4   The US government

It may seem odd to include the US government among a list of customers. But in regard to energy, it is the largest single consumer in the US. Perhaps this is what motivated President Obama in 2009 to issue Executive Order (EO) 13514, Federal Leadership in Environmental, Energy, and Economic Performance, which states the following:

> *In order to create a clean energy economy that will increase our Nation's*
> *prosperity, promote energy security, protect the interests of taxpayers, and*
> *safeguard the health of our environment, the Federal Government must lead by*
> *example.*[16]

Implementation of this order mandated aggressive goals for the reduction of GHG resulting from energy usage at federal facilities. Although the GHG goals varied by federal agency, for the DOE, their goal was a 28% reduction in GHG emissions by

2020 relative to a 2008 baseline. The EO catalyzed considerable activity among federal facility operators across the country to develop an energy use strategy that could achieve the mandated reduction in GHG emissions while meeting expected energy demands.

8.4.1 Non-DOD federal facilities

Within the federal government, the Department of Defense (DOD) is the largest energy consumer. Among the non-DOD agencies, the highest energy users are the DOE national laboratories, which operate several mission-critical research facilities such as high-powered accelerators and supercomputers. In response to EO 13514, the DOE reviewed the GHG emission data for each of its laboratories and requested each site to prepare a plan on how it would achieve the mandated level of GHG reduction by 2020. Table 8.3 lists the laboratories with the highest emissions, including laboratories managed by the National Nuclear Security Administration (NNSA), the DOE Office of Environmental Management (EM), and the DOE Office of Science (SC).[17] Of the DOE/SC laboratories, ORNL was the largest energy consumer and consequently the most grievous in terms of GHG emissions. This dubious honor made ORNL the natural target for considerable scrutiny by DOE/SC headquarters staff.

Coincidentally, at the same time that EO 13514 was issued, the DOE's Office of Nuclear Energy (DOE/NE) was developing their new SMR program, which had two key components: accelerate the deployment of near-term SMR designs and conduct R&D that supports the development of advanced SMR technologies. At the instigation of ORNL, DOE/NE's interest in promoting SMRs came together with DOE/SC's interest in meeting EO 13514 mandates to yield an obvious solution: build an SMR at ORNL to support its research mission while achieving a net-zero GHG emission

Table 8.3 Baseline GHG emissions for the highest ranking DOE/ NNSA laboratories[17]

Site	State	Program	GHG emissions (MtCO$_2$ equivalent)
Savannah River	SC	EM	515,779
Los Alamos National Laboratory	NM	NNSA	410,896
Y12 National Security Complex	TN	NNSA	272,560
Sandia National Laboratories	NM	NNSA	266,087
ORNL	TN	SC	258,597
Fermi Lab	IL	SC	252,791
Portsmouth	OH	EM	203,260
Argonne National Laboratory	IL	SC	183,510
Lawrence Livermore National Laboratory	CA	NNSA	123,506
Brookhaven National Laboratory	NY	SC	123,273

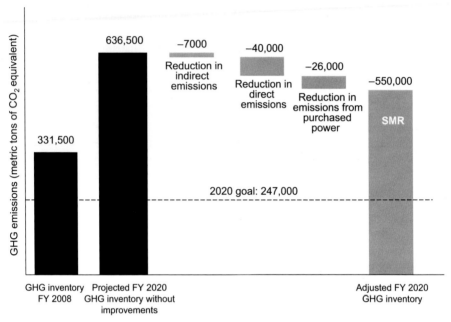

Figure 8.2 Proposed path to achieving zero GHG emissions at ORNL by 2020.[18]

facility. What resulted was the development of a compelling business case for an SMR at ORNL—envisaged to be the first SMR to be deployed in the US.[18]

The underlying basis of the ORNL SMR business case is shown in Figure 8.2. The bar on the far left gives the baseline 2008 GHG emissions for ORNL, and the dotted horizontal line shows the 2020 emissions goal based on a 28% reduction in the 2008 value. The second bar from the left shows the projected GHG emissions for ORNL in 2020 based on anticipated demand growth and assuming no measures were taken to reduce emissions. Incidentally, nearly all of the growth in electricity demand was projected to be due to supercomputing hardware, which was estimated to require at least 75 MWe. The three middle bars in Figure 8.2 show the potential reductions in various categories of energy usage and assume relatively aggressive improvements in ORNL's energy efficiency as well as the local utility's shift to cleaner electricity generation. Finally, the right-hand bar shows the reduction in GHG emissions that would be achieved by a single 125 MWe SMR dedicated to the ORNL facility. Not only does the SMR allow ORNL to achieve a net-zero GHG emission level, but it is the only tractable solution for achieving the 2020 emission goal mandated by EO 13514. This analysis resulted in the formation of a collaboration among ORNL, TVA, Babcock & Wilcox, and the DOE to pursue the deployment of an mPower SMR on the Clinch River site, which is immediately adjacent to the ORNL reservation. The ORNL SMR business case also became a model for other federal facilities to consider in meeting their future energy needs consistent with EO 13514.

A few years after the DOE/NE SMR program was funded, the DOE tasked ORNL with systematically evaluating all federal facilities for potentially building an SMR

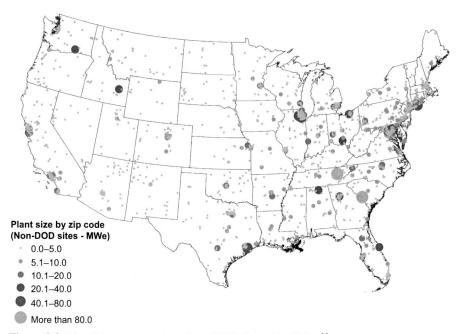

Figure 8.3 Electricity consumption of non-DOD federal facilities.[19]

onsite to support their missions. Using the OR-SAGE site evaluation tool, ORNL concluded that most of the individual non-DOD facilities had power consumptions below
the threshold of current SMR designs.[19] Figure 8.3 provides a map of the location of
all non-DOD federal facilities and an indication of their power usage, which shows
that the vast majority of the roughly 4,800 individual facilities consume less than
80 MWe each. Only the ORNL and the Savannah River Site exceeded 80 MWe.

However, by aggregating nearby facilities, ORNL was able to identify several clusters of federal facilities that appear to be well suited for being powered by an SMR.
Specifically, they identified 13 clusters of facilities that had power demands greater
than 200 MWe, 8 of which satisfied several evaluation criteria for assessing the site
suitability for an SMR. These are listed in Table 8.4. To my knowledge, only the East
Tennessee location is actively moving forward with a potential SMR deployment, but
all eight locations provide promising opportunities for the government to step forward
as first movers for the utilization of clean power from SMRs.

8.4.2 DOD facilities

While the DOE has been reasonably aggressive at evaluating opportunities to be the
"first mover" for domestically sited SMRs at their national laboratories, the DOD
could be characterized as a cautious observer and potential second-mover. Several
engagements have been made between the DOE and DOD on the topic since 2009;
however, no sustained champion has emerged from within the DOD leadership, and
no individual DOD facility has pursued deploying an SMR on its site to the extent of

Table 8.4 **Most promising federal facility clusters for potential SMR deployment**[19]

Location	Power demand (MWe)
Virginia Peninsula (Hampton Roads area)	369
South Carolina (Savannah River)	337
Florida Panhandle	305
Southcentral Texas	252
Denver–Colorado Springs, Colorado	238
East Tennessee (Oak Ridge Reservation)	234
Southwest Oklahoma–North Texas	219
Western Ohio	206

ORNL and other national laboratories. In late 2009, a meeting between the DOE and DOD, with NRC participating, initiated the dialogue of potentially powering domestic DOD bases using SMRs. The multiagency collaboration resulted in a feasibility study conducted by the Center for Naval Analyses (CNA).[20]

The CNA study concluded that SMRs are a feasible option for DOD installations and that they would help the DOD to meet several important goals, including the EO 13514 goals for the reduction of GHG emissions. More importantly, they would provide energy security for critical national security missions. The smaller size of SMRs, the higher level of reliability afforded by module redundancy in multimodule plants, and their enhanced safety appear to be the primary SMR features of interest to the DOD. The study also articulated three primary concerns: (1) the risk (financial obligation) of being an early adopter of new technology, (2) the generally small power demand of many domestic installations, and (3) the suitability of DOD installations for the siting of a nuclear plant. Shortly after the CNA study, the threat of budget sequestration for the DOD and the potential for base closures exacerbated the DOD's caution in embarking on an SMR deployment strategy for domestic bases.

The first concern, that is, the risk of being an early adopter, is quite common. I have heard several SMR vendors say that they have many second customers but are still waiting for the first customer. Unquestionably, the acceptance of innovation and new technologies in the nuclear industry is a huge challenge. The other two concerns documented in the CNA report are more tangible. Figure 8.4 shows the distribution of average power usage by domestic DOD facilities as reported by the CNA.[20] Nearly 94% of the facilities have a power usage less than 50 MWe, which corresponds to the smallest of the SMRs currently being commercialized in the US. However, many of the facilities are in close proximity to other federal facilities, which spurred the ORNL OR-SAGE researchers to evaluate aggregated federal sites, as discussed earlier.[19] Several of the eight federal cluster locations listed in Table 8.4 include DOD installations, either solely or in combination with non-DOD facilities. Furthermore, all eight cluster locations satisfied the several SMR site evaluation criteria used in the ORNL analysis.

On March 19, 2015, President Obama issued a new EO to "improve environmental performance and Federal sustainability." The order, entitled "Planning for Federal

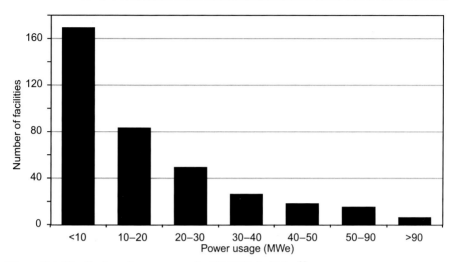

Figure 8.4 Distribution of power usage by US DOD facilities.[20]

Sustainability in the Next Decade," calls on federal facilities to reduce energy usage and to increase their percentage of renewable and alternative energy sources, which explicitly include SMRs.[21] The order further asserts that "pursuing clean sources of energy will improve energy and water security, while ensuring that Federal facilities will continue to meet mission requirements and lead by example." Perhaps this new EO will further motivate both DOD and non-DOD installations to pursue SMRs.

In summary, the natural courting ritual between sellers and buyers is a long, arduous, and highly visible process in the nuclear business. Bringing a new reactor design to market can take more than a decade and consume a billion dollars. Also, there are many commercial, regulatory, political, and social gates to clear. Still, there are many customers who have come forward with significant interest in SMRs. They are generally well informed and quick to understand how the specific features of SMRs can help meet their energy needs. Their decision is not an easy one. There are many challenges remaining to fully deploy new SMR designs, one of which is an almost deafening level of media chatter, pro and con, which can distract even the most committed customer. In the next chapter, I review many of these challenges. However, with these challenges come many opportunities—opportunities to design, license, manufacture, and operate nuclear power plants in improved and innovative ways.

References

1. *Guide to electric power in Ghana*. University of Ghana; July 2005.
2. *Assessing policy options for increasing the use of renewable energy for sustainable development: modelling energy scenarios for Ghana*. New York, NY: United Nations; 2006.
3. *Common user considerations (CUC) by developing countries for future nuclear energy systems: report of stage 1*. International Atomic Energy Agency; 2009. NP-T-2.1.

4. Black G, Black MA, Solan D, Shropshire D. Carbon free energy development and the role of small modular reactors: a review and decision framework for deployment in developing countries. *Renewable Sustainable Energy Rev* 2015;**43**:83–94.

5. Kessides IN, Kuznetsov V. Small modular reactors for enhancing energy security in developing countries. *Sustainability* 2012;**4**:1806–32. http://dx.doi.org/10.3390/su4081806.

6. *Milestones in the development of a national infrastructure for nuclear power.* International Atomic Energy Agency; 2007. NG-G-3.1.

7. *Small nuclear reactors for Oak Ridge?* OakRidger.com; November 11, 2010. Available at: www.oakridger.com/news/x684363339/Mini-nuclear-reactor-for-Oak-Ridge.html.

8. Celebi M, Graves F, Russell C. *Potential coal plant retirements: 2012 update.* The Brattle Group; 2012.

9. Annual Electric Generator Report, Form EIA-860, Energy Information Administration, Available at: www.eia.gov/electricity/data/eia860.

10. Banks GD. *Why utilities want small modular reactors.* Center for Strategic and International Studies; August 13, 2013. Available at: http://csis.org.

11. Belles RJ, et al. *Evaluation of suitability of selected set of coal plant sites for repowering with small modular reactors.* Oak Ridge National Laboratory; 2013. ORNL/TM-2013/109.

12. Wardell J. *Cities may turn to new forms of nuclear power.* The Davis Clipper; November 21, 2014. Available at: http://davisclipper.com/view/full_story/26126001/article-Cities-may-turn-to-new-forms-of-nuclear-power.html.

13. Atkinson D. *Why energy northwest is interested in SMRs.* NEI Nuclear Notes; April 10, 2014. Available at: http://neinuclearnotes.blogspot.com/2014/04/why-energy-northwest-is-interested-in.html.

14. *Overview of energy development opportunities for Wyoming.* Idaho National Laboratory; 2012. INL/EXT-12-27626.

15. *Preliminary feasibility of value-added products from cogeneration and hybrid energy systems in Wyoming.* Idaho National Laboratory; 2012. INL/EXT-12-27249.

16. *Federal leadership in environmental, energy and economic performance,* executive order 13514. *Fed Regist* October 8, 2009;**74**(194):52117–27.

17. *Meeting the challenge of executive order 13514.* US Department of Energy; January 29, 2010. presentation to the Senior Sustainability Steering Committee.

18. Ingersoll D. *Case study: potential SMR deployment at a US Government R&D facility.* Oak Ridge National Laboratory; October 11, 2011. presented at the INPRO Dialog Forum on Common User Considerations for SMR.

19. Belles RJ, Mays GT, Omitaomu OA, Poore WP. *Identification of selected areas to support federal clean energy goals using small modular reactors.* Oak Ridge National Laboratory; 2013. ORNL/TM-2013/578.

20. King M, Huntzinger L, Nguyen T. *Feasibility of nuclear power on US Military installations.* Center for Naval Analyses; March 2011. CRM D0023932.A5/2REV.

21. Executive Order. *Planning for federal sustainability in the next decade.* Office of the Whitehouse; March 19, 2015.

Getting to the finish line: deployment challenges and opportunities

Earlier attempts in history to deploy commercial small modular reactors (SMRs) did not succeed, although I conclude that it was not because of any fault in the basic premise of SMRs but rather because of externalities that shifted momentum away from nuclear power in general. As evidence of this, with each renewed interest in nuclear power, so comes renewed interest in SMRs. We appear to be closer now than ever before in getting SMRs into the market, yet there are several remaining challenges—gates to be cleared and hurdles to be resolved—in order to finally cross the finish line.

In this chapter I offer my perspective on the remaining challenges and associated opportunities for the deployment of SMRs. I have organized them into three primary categories: technical, institutional, and social. I finish with a brief discussion of the role of the US government—past, present, and future—in helping to meet those challenges and enable SMRs to finally secure their place in the US and global energy future.

Before embarking on a discussion of challenges, I should clarify my use of the word "challenge." I have observed a broad spectrum of interpretations for the meaning of "challenge." Opponents like to define challenges as insurmountable obstacles, that is, impediments that will prevent the ultimate deployment of SMRs. Researchers, a community with whom I have spent most of my career, make their livelihoods out of solving challenges, so they define them as funding opportunities. Now that I work for a commercial SMR vendor, I observe firsthand that engineers define challenges merely as items on their task lists. For the purposes of this chapter, I prefer to interpret the word more in line with the Encarta dictionary: "stimulating test of abilities." This reflects both the yet-to-be-resolved character of the word as well as the opportunity to improve and innovate. The challenges still remaining for SMR deployment will require a concerted effort to resolve them but also provide abundant opportunities—a chance to deploy new technologies in new ways with the potential to change how we think about nuclear power. So in this chapter, I comingle both challenges and opportunities.

9.1 Technical challenges and opportunities

The first and most obvious challenge for SMR designers is to finish their designs. It is not required to have a fully complete design in order to license the design, or even start construction, but it is a really good idea. Most agree that the huge cost overruns and construction delays during the original build-out of our nuclear fleet were due substantially to "just in time" designing. Unfortunately, not all vendors have learned

Small Modular Reactors. http://dx.doi.org/10.1016/B978-0-08-100252-0.00009-4

this lesson, and some of the current new builds are experiencing similar overruns and delays for similar reasons. The most outstanding example of this is the construction of Areva's first EPR plant in Finland, which is now several years behind schedule and billions of dollars over budget.[1] So the first technical challenge is to get the job done, that is, complete the design and engineering before starting construction. If SMRs are to succeed, it is imperative that vendors not succumb to investor or customer pressure and push new technologies and engineering into the marketplace without adequate maturation and validation.

9.1.1 Technical challenges

A big part of getting the job done is ensuring that the underlying technologies are sufficiently mature to allow the engineering to be completed. The current fleet of large light-water reactors (LWRs) has set a high performance standard, and new plants will be expected to meet this high standard, if not surpass it. New process heat application markets enabled by SMRs will also demand reliable operations in addition to safety and cost effectiveness. More than 60 SMR concepts are in various stages of development worldwide, and most have been sufficiently studied to validate the fundamental physics of the concept. However, far fewer concepts have been engineered to the point that the design's performance and its ability to deliver a commercially viable product, that is, electricity or heat, can be reasonably assured.

SMRs that are based on LWR technology clearly have the advantage in terms of technology maturity. According to a 2007 assessment conducted at Oak Ridge National Laboratory, more than 20,000 reactor years of operational experience have been gained with LWRs worldwide, including both commercial power plants and naval propulsion units.[2] This compares to less than 2,000 reactor years for gas-cooled reactors, roughly 300 reactor years for sodium-cooled reactors, 80 reactor years for lead-bismuth-cooled reactors, and four reactor years for salt-cooled reactors. As reviewed in Chapter 4, the reactor coolant typically impacts the entire reactor system in terms of the choice of fuel, materials of construction, and components. Hence, in the case of non-LWR SMR designs, significant technology maturation may be needed for commercialization compared to water-cooled designs.

Technical challenges for non-LWR SMRs are generally not specific to their smallness but rather are a reflection of their fundamental physics or operational parameters. For instance, the fuel may utilize high-energy fission reactions or low-energy fission reactions, which impacts the rate and type of radiation-induced damage in surrounding materials. Also, the reactor coolant may operate at low, medium, or high temperatures, which can impact the choice of materials throughout the reactor system. In most cases, testing and qualification will be needed for new materials and fuels in addition to new components and systems relative to existing plants. Even fuels that are incrementally different from traditional nuclear fuel can require 10–12 years and a billion-dollar investment to fully qualify them for use in a commercial plant. The long time line results from a complex, sequential testing protocol: (1) irradiation of test specimens (can require multiple years to accumulate sufficient exposure), (2) postirradiation examination and analysis of the test specimens, (3) fabrication of a prototypical fuel element, and finally (4) performance testing of the prototype fuel element. Material and fuel qualification will be especially important for SMR designs that are intended to have very long fuel cycle lengths—some designs claim cycle lengths as much as 10–15 times longer than current LWR fuels. This means that the fuel material and the protective cladding must maintain their integrity despite a tenfold increase in radiation-induced damage.

Most designers of LWR-based SMRs have deliberately chosen to minimize technology risks in order to expedite the development and licensing processes and minimize testing and qualification costs for first-of-a-kind components and systems. However, even with LWR-based designs, some technology development is needed because of the nontraditional design configurations. For example, the integral primary system configuration that many SMR vendors use may introduce some nonconventional components such as helical coil steam generators, internal control rod drive mechanisms, and internal coolant pumps. These new components must be fully engineered, tested, and qualified for the operational environment in order to meet the expectations of both the regulator and the customer.

Additional developments in sensors, instrumentation, and control systems may be needed for some SMR designs. For example, the lack of external primary coolant loops in integral SMR designs means that conventional measurements of coolant flow and heat balance are not possible, and new approaches for in-vessel measurements must be developed and qualified. Also, operation of SMRs in more remote locations encourages the implementation of additional sensors and instrumentation for online plant health monitoring, diagnostics, and prognostics. Depending on the application of the SMR, especially for cogeneration applications, new control systems may need to be developed to appropriately manage multiproduct load balancing.

Most SMR vendors fully understand these requirements and have extensive testing and validation programs underway. Several of these testing programs are reviewed in Chapters 14–19 of the *Handbook of Small Modular Nuclear Reactors,* which provide overviews of the supporting research and development programs within several countries that are developing new SMR designs.[3] In the case of NuScale Power, they are well into a multiyear testing program that will eventually provide validation of the performance, environmental qualification, and manufacturability of all first-of-a-kind components.[4] The path is well known; it just needs to be traveled.

In addition to maturing and qualifying the hardware, there is the need to develop and validate computational analysis methods that are used to predict the safety and operational performance of the reactor components and systems. Many industry-qualified codes exist, especially for LWR-based reactor designs, but differences in design configuration can create new data and validation needs. For example, some SMRs use natural circulation flow of the primary coolant for normal operation. This is a significant departure from existing large nuclear plants and requires comprehensive validation of the methods used to predict thermal-hydraulic performance of natural circulation flow for all operational conditions. Even if the operating conditions of the reactor are more benign than existing experience, such as lower coolant pressure or core power density, validation of data used by the analysis codes must be performed for those conditions.

Experimental validation of system thermal-hydraulic performance is best accomplished using a test facility that models the entire primary reactor system. New reactor designs are frequently preceded by a prototypical low-power nuclear reactor or a scaled non-nuclear system simulator. The latter is more common for LWR-based designs since only the overall system performance requires testing rather than the underlying technology. Both NuScale Power and Generation mPower have constructed scaled, electrically-heated simulators to validate the overall performance of their designs. Other SMR vendors have also built similar test facilities or have plans to do so. In the case of the CAREM SMR developed in Argentina, a 25 MWe prototype reactor is being constructed in anticipation of an eventual 100–150 MWe commercial SMR.[5]

An important computational method for analyzing the safety and reliability of a plant, especially for assessing the likelihood of a serious accident occurring, is called probabilistic risk assessment (PRA). The methodology is well developed for single reactor units but is not typically used to account for cross-unit interactions, that is, propagation of an accident sequence from one reactor unit to adjacent units. The accident at the Fukushima Daiichi plant in Japan that involved four of the six units demonstrated that cross-unit interactions can happen, for example, hydrogen from one unit leaked into an adjacent unit, causing an explosion there. Cross-unit interactions need to be evaluated as part of the total plant safety assessment. SMR designers have the benefit of learning from the Fukushima experience and also accept from the outset that multimodule plant designs will require multimodule analyses. Existing PRA codes can be adapted to analyze multimodule effects; however, it is useful and desirable to extend PRA methods to more readily represent multimodule interactions.

Another technical challenge, although it might also qualify as a social challenge, is the need to vigilantly guard against traditional engineering mindsets during detailed finalization of the SMR design. For an SMR to be economically successful, it must maximize the economies of small, which depend substantially on design simplicity. This requires a new discipline for design engineers and their leadership to not allow unnecessary complexity to creep in as design challenges are encountered. Admiral Hyman Rickover, father of the US Nuclear Navy, is widely quoted for his comparison of a paper reactor to an operating reactor. Among several differences, he stated that paper reactors are always simple while operating reactors are always complex. While I agree with this generality, it need not always be true. A counterexample is that when the NuScale electrical engineers were faced with the prospects of needing a substantial number of safety-grade batteries in their 12-module SMR plant, they designed a system that now requires no safety-grade batteries to safely shut down and cool the reactors. This is a great example not only of maintaining engineering simplicity but also of turning challenge into opportunity.

9.1.2 Technical opportunities

The fact that most SMRs are still in the design phase opens up the opportunity to apply new technologies not only for the designs themselves but also for the design process. For instance, high-performance computational simulation has advanced dramatically in other fields since the time that existing nuclear plants were designed. There is a tendency to perpetuate the previous nuclear analysis methods, partly because of the cost of developing new methods at the high level of quality assurance rigor required for commercial application, and also because of the familiarity and acceptance of the established methods by the regulator. However, new simulation methods offer many benefits. They incorporate higher fidelity modeling of the underlying phenomenology than the empirical-based codes of the past, which means that they can be more reliably extended to the analysis of new design configurations. Also, they are optimized for the architectures of modern supercomputers, which means that design analyses can be run in a small fraction of the time compared to the older methods, thus facilitating more comprehensive analyses.

Fortunately, there have been sustained investments in recent years by the Department of Energy (DOE) in applying advanced simulation methods to nuclear power. The Nuclear Energy Advanced Modeling and Simulation program and a DOE Innovation Hub, referred to as the Consortium for Advanced Simulation of Light-Water Reactors, are examples of programs focused on developing state-of-the-art simulation methods that can improve the accuracy and speed of nuclear analyses. These new methods offer the potential to expedite the SMR design process and increase confidence in the predicted plant performance. In turn, existing SMR system simulators, and eventually the first SMR units, offer an opportunity to provide method developers with valuable validation data.

Related to advanced simulation is the opportunity to apply advanced, three-dimensional visualization methods to reactor system design and operation. Fully immersive virtual environments are now available that allow the design engineer to move within the design model to verify component placement and size and to check for spatial conflicts. During operation, three-dimensional virtual environments can facilitate maintenance training and ensure suitable space and pathways for component inspection, repair, or replacement. Similar opportunities exist to improve the manufacturing of reactor modules using techniques for generating three-dimensional models of the manufactured system and automated methods for comparing the as-built module to the original design model.

A more subtle opportunity enabled by the fundamental nature of SMRs is the potential to introduce new technologies at a faster pace with lower financial risk than has been the norm in the highly conservative nuclear industry. Specifically, there are several examples of modern materials that are stronger, cheaper, and more enduring than traditional nuclear plant materials, which were mostly developed during the 1960s and 1970s. The same is true for advanced manufacturing and fabrication technologies such as components fabricated by powder metallurgy processes and advanced welding or cladding techniques. Many designers forego introducing new materials or manufacturing techniques into their designs because of the licensing requirements to sufficiently prove that the new technology is safe for service, including the additional time needed to obtain acceptance by relevant standards committees. In principle, however, the multimodule paradigm of SMRs and the substantially lower fractional cost of a single module in the plant could encourage earlier introduction of new technologies into the plant as new modules are added or old ones replaced.

9.2   Institutional challenges and opportunities

I have spent most of my career studying nuclear reactor technology, so I am very familiar and comfortable with technology challenges. I am also confident regarding a successful outcome. I am less knowledgeable and much less comfortable with the path needed to overcome several institutional challenges, which I categorize as regulatory, legal, and business issues. Technology challenges are probably the "long pole in the tent" for non-LWR SMRs because of the high cost and lengthy processes needed to develop, test, and qualify new technologies for use in a commercial nuclear plant. On the other hand, institutional issues may be the dominant challenges for SMRs that are based on LWR technology primarily because these designs are further along the deployment path than the non-LWR designs and will hit the institutional issues first. In effect, they will blaze the trail through these challenges for the new technologies that may follow.

9.2.1   US regulatory challenges

Globally, three SMRs have received regulatory approval of the designs in their respective countries: the SMART reactor developed by the Republic of Korea, the KLT-40S that is being constructed in the Russian Federation, and the HTR-PM that is being constructed in China. The US Nuclear Regulatory Commission (NRC) began to recognize in the early 2000s that considerable activity among SMR vendors would likely result in the submittal of new license applications and that changes to the current licensing framework might be required. Westinghouse was the first to approach NRC for preapplication activities regarding the IRIS design, followed by Toshiba for the potential deployment of their 4S design in Alaska. Both of these engagements were later suspended, but other SMR designers such as NuScale Power and Generation mPower initiated prelicensing activities. In 2010, the NRC issued a paper detailing more than a dozen regulatory issues that may need resolution in order to successfully license a new SMR design.[6] The potential issues were described in a staff report, designated SECY-10-0034, and are summarized in Table 9.1.

Although the list of issues might appear daunting, the NRC began immediately to work through them in a prioritized fashion and developed several follow-up staff position papers, which are also listed in Table 9.1 in the column labeled "Follow-up SECY." In addition, separate industry efforts were initiated by special committees within two different organizations, the American Nuclear Society (ANS) and the Nuclear Energy Institute (NEI), to identify potential regulatory hurdles and propose solutions. Although there is not a one-for-one mapping of issues identified by the NRC and those identified by ANS and NEI, many are the same or similar. Those NRC-identified issues for which ANS or NEI position papers have been drafted are included in Table 9.1 in their respective columns.

One licensing issue that was not included in SECY-10-0034 but identified by both the ANS and NEI working groups relates to the identification and implementation of a process referred to as ITAAC, which stands for "inspections, tests, analyses, and acceptance criteria." The ITAAC are an aspect of the 10 CFR, Part 52 licensing framework that ensures that the nuclear plant is built exactly according to the design that was reviewed and approved by the NRC. This process is being used for the new AP-1000 plants that

Table 9.1 **Potential SMR licensing issues identified by the NRC[6]**

Issue in SECY-10-0034	Follow-up SECY	ANS paper	NEI paper
Licensing process			
Licensing of prototype reactors	11-0112	Yes	
License structure for multimodule plants	11-0079		Yes
Manufacturing license requirements		Yes	
Design requirements			
Implementation of defense-in-depth			
Use of probabilistic risk methods	11-0156	Yes	
Source term, dose calculations, and siting			Yes
Key component and system design issues			
Operational requirements			
Requirements for operator staffing	11-0098	Yes	Yes
Operational programs	11-0112		
Installation of additional modules	11-0112		
Industrial facilities for process heat	11-0112	Yes	
Security and safeguard requirements	11-0184	Yes	Yes
Aircraft impact assessment	11-0112		
Offsite emergency planning requirements	11-0152	Yes	Yes
Financial implications			
Annual fee structure for multimodule plants		Yes	Yes
Insurance and liability	11-0178	Yes	Yes
Decommissioning funding	11-0181	Yes	Yes

are being constructed in Georgia and South Carolina, and in principle, the same process will apply to SMR construction. However, the greater fraction of factory fabrication that will occur for most SMR designs may result in new issues regarding the implementation of factory-based ITAAC and potential receipt verification after shipment of the reactor module to the plant site. On the plus side, SMRs may provide an opportunity to simplify and standardize the ITAAC. This opportunity is being pursued by the NEI SMR Working Group, which is working to develop standardized terminology and classes of ITAAC that will help to remove uncertainties and ambiguities from the ITAAC process.[7]

In addition to the position papers developed by the ANS and NEI, individual vendors are engaging the NRC directly in prelicensing activities to resolve as many of the regulatory issues as possible prior to the submission of an application. At least two SMR vendors, Generation mPower and NuScale Power, are working with NRC to develop a design-specific review standard to improve the efficiency of the regulatory review after the license application has been submitted. A design-specific review standard is basically a blueprint of how the license application will be reviewed by the NRC and can account for differences in the SMR design relative to large traditional plants or even other SMRs.

As is the case with technical challenges, most regulatory challenges are well known, and the path for resolving them is generally apparent. Many organizations such as the NEI are actively engaged and working together with the NRC to resolve these issues in advance of license application submittals. However, the burden of proof is always on the license applicants to convincingly demonstrate the safety of their SMR design and justify any changes or exceptions to existing regulations without compromising the safety standards established by the NRC. A major uncertainty is how the regulator will respond to the differences between SMR designs compared to more familiar large plant designs and the robustness of the regulations (and the regulator) to adapt to these differences.

Emergency planning is a highly visible example of an issue for SMRs that will require thorough substantiation by the applicant and adaptability by the regulator. All currently operating commercial power plants in the US operate with a 10-mile radius for their emergency planning zone (EPZ) for plume exposure. The regulations allow for a smaller radius of the EPZ if justified by a lower plant risk. Many SMRs offer a substantial reduction in plant risk due to the slower rate of accident progression, the reduced likelihood of fuel damage in the case of an accident, a smaller amount of radiological hazard in the core, and additional engineered barriers to limit radioactivity release in the case of fuel damage. The challenge for the SMR vendor and the plant owner is to prove that a simplified emergency response plan and reduced EPZ do not increase risk to the public relative to current practice, and the challenge for the regulator is to allow regulatory relief from the expenses associated with the larger EPZ if the risk case is convincingly substantiated. There is precedence for this since the NRC granted a reduced EPZ for three former nuclear plants, although none are currently operating.[8]

A reduced EPZ is important to SMR deployment for a number of reasons. First, there is a desire by utilities to use SMRs as a one-for-one replacement for retiring coal plants, some of which may be close to population centers. Also, the use of SMRs for process heat applications would be enhanced by a reduced EPZ and exclusion zone since they could facilitate shorter heat transport lines between the nuclear plant and the industrial facility. Finally, a significant initial and annual cost savings would be realized. To facilitate the process of substantiating their case by SMR vendors and owners, the NEI prepared a white paper that outlines a methodology and criteria that can be used to establish a technical basis for a design-specific EPZ.[9]

In practice, the decision on EPZ will be very complex and involve a large community of stakeholders such as local and state governments, emergency responders, and the local population. However, many experts believe that there is ample reason to justify a substantial reduction in the EPZ and simplification of emergency management for many SMRs. In terms of the designs with which I am most familiar, they appear to have a high level of resilience, including a likelihood of fuel damage that is orders of magnitude less than current plants, additional physical barriers to reduce radioactive releases, and little or no dependence on active human or engineered systems to maintain a safe condition of the plants. Another factor, sometimes overlooked in the EPZ debate, is that local emergency response capabilities have improved dramatically since the September 11, 2001, terrorist attacks on the World Trade Center and the Pentagon, and are now better equipped to respond to all types of incidents, including industrial accidents.

A trend in nuclear plant regulation that will help increase regulatory adaptability is the use of risk-informed decision-making processes. The underlying principle is that the regulatory requirements should be commensurate with the plant risk, which is a combination of the likelihood of failure and the consequences from a failure. This principle applies to design features, such as the need for component redundancies or backup systems, and operational procedures, such as in-service inspection require-ments. The substantially lower risk factors in several SMR designs should provide a compelling demonstration of the efficiency and cost improvements that can be gained by the appropriate application of risk-informed regulation. Under the leadership of NRC Commissioner George Apostolakis, the NRC drafted a proposed framework for incorporating risk insights into the review of license applications for SMRs.[10] The framework allows for a graded approach to the review of SMR applications that puts a high level of detailed review on structures, systems, and components that are safety related or have safety significance and a lower level of detail on nonsafety-related design features. Besides offering a more efficient review process, the framework helps to move the NRC toward a more adaptable performance-based review process rather than the former checklist-based deterministic approach.

Another licensing issue that has received considerable industry press is regarding plant staffing for SMRs, especially the size of the security force and the number of reactor operators in the control room. Most SMR designers are incorporating intrinsic features into their designs, such as below-grade placement of safety systems and lim-ited access routes to nuclear fuel, in order to achieve a more optimized on-site security force. Changes in control room staff size impact only a few SMRs since many designs incorporate only one or two modules in a single plant. In the US, only the NuScale SMR exceeds the current practice of two reactor units per control room. NuScale has indicated their intent to request an exemption to the current limit since they plan to operate up to 12 modules in a single control room. The path for requesting the exemp-tion was established in 2005 in anticipation of a license application for a multimodule pebble-bed modular reactor (PBMR).[11] However, a license was never filed for the PBMR, so it is a path untraveled.

A licensing challenge for the owner is whether to use the former 10 CFR, Part 50 licensing framework or the newer "one step" Part 52 framework. In the Part 50 process, the owner must first obtain a license to construct the nuclear plant, and then once constructed, it must obtain a license to operate the plant. This two-part process proved to be costly during the original build-out of plants in the 1960s and 1970s due to delays in obtaining the operating license while the bright, shiny, and expensive plant sat idle. Under Part 52, the owner applies for a combined construction and operating license before sinking his or her life savings into constructing the plant. While this one-step approach seems less risky, the challenge comes from the potential for design changes during construction, either as a result of an incomplete design at the time of construction or changes that are identified as a result of trying to build the plant. The flip side of this is that because in-construction design changes are more cumbersome with Part 52, it effectively promotes design standardization—another important lesson from the early plant build-out.

The Tennessee Valley Authority (TVA) initially indicated that they intended to use the Part 50 approach for their deployment of an mPower SMR on a site in Tennessee, citing as justification that the older process is more adaptable to minor design changes during construction. After Generation mPower reduced their pace of development, TVA decided to discontinue their pursuit of a Part 50 construction permit and instead proceed with obtaining a Part 52 early site permit enveloping a broader range of potential SMR designs. Likewise, the Utah Associated Municipal Power Systems announced in late 2014 that they plan to use Part 52 for the deployment of the first NuScale SMR plant in Idaho. So it appears that only the Part 52 process may be used for the first SMR deployments, but the debate on which process is better suited for constructing first-of-a-kind designs in the US continues. In terms of the international market, the Part 52 process has the clear advantage to the vendor of achieving an NRC Design Certification, which is a major competitive advantage.

A class of issues that cuts across technical and regulatory challenges is the suitability of existing codes and standards relevant to the materials and components to be used in SMRs. This will not be a significant hurdle for most LWR-based SMR designs since the designers have deliberately selected materials and components that meet current codes and regulations. In addition, using an integral design configuration eliminates large pipes and large pressure vessel penetrations from the design, which consequentially eliminates several code issues associated with those design features. On the other hand, the elimination of large-bore piping in integral systems may change the safety function of small-bore piping, which creates an ambiguity in terms of the applicability of relevant codes and standards for small pipes and penetrations.

Changes or additions to existing codes and standards are almost certain to be needed for non-LWR SMR designs. Changes may also be advantageous even for LWR-based designs to facilitate operational improvements such as extended fuel cycles and reduced in-service inspections. The Nuclear Energy Standards Coordination Collaborative (NESCC) was formed in 2009 to review existing nuclear industry codes and standards to ensure that they were current and reflected new technology. Its purpose has been expanded to help coordinate the identification and development of standards needed by new plant designs, including SMRs. Both the NRC and DOE participate in the NESCC, which is managed by the American National Standards Institute and the National Institute of Standards and Technology. The group serves to coordinate codes and standards requirements of the nuclear community with several standards development organizations.

Although there are several regulatory issues to work through, including some first encounters due to SMR design differences, the NRC and the industry are cooperatively engaged and working through the remaining challenges. In a position paper issued by the NRC, they confidently state that they are prepared to review the application of SMR designs and have developed a 39-month review schedule.[12] This assertion is contingent on sufficient prelicensing engagement by the applicant and the completeness of the application; however, it is an encouraging indication that US-based SMRs may actually get to the finish line this time around.

9.2.2 International licensing

The widespread interest in SMRs globally and their features that facilitate broader deployment opportunities have catalyzed multiple groups to propose and pursue international harmonization and standardization of SMR licensing. Prior to the current escalation of interest in SMRs, a multinational design evaluation program (MDEP) was formed in 2006, led by the Nuclear Energy Agency, to allow regulators from multiple countries to share their experience in licensing new plant designs.[13] Initially, the program has been focused on several large plant designs of interest to multiple countries, but the organization's policy framework allows it to be easily extended to include SMR designs. An SMR Working Group was formed within the Cooperation in Reactor Design Evaluation and Licensing (CORDEL) Working Group of the World Nuclear Association to coordinate international activities related to SMRs, including standardized certification and licensing.[14]

In July 2013, the 6th Dialogue Forum for the International Project on Innovative Nuclear Reactors and Fuel Cycles was held at the International Atomic Energy Agency (IAEA) and focused on international safety and licensing standards for SMRs.[15] The 120 representatives from 37 countries and 4 international organizations participated in several breakout and plenary sessions to review and evaluate the suitability of IAEA safety standards and guidelines relative to SMR designs being developed worldwide. The preliminary conclusions were that most of the IAEA safety standards were applicable to SMRs, although the agency's safety guidelines require further evaluation. A major outcome of the meeting was a broad consensus on the value of establishing an SMR Regulator's Forum, which can serve as a focal point for the harmonization of SMR licensing processes and provide effective coordination with other related groups such as MDEP and CORDEL.

To be sure, there are many challenges in achieving true harmonization of nuclear plant licensing. National sovereignty, variations in legal requirements, and cultural differences will complicate the harmonization process. However, SMRs may provide the catalyst to more aggressively pursue this ambitious goal. Because of this, there has been a steady increase in the interest and activities to develop frameworks for achieving international licensing of SMRs. Kristiina Söderholm, who leads the CORDEL effort on SMR licensing harmonization, published a paper based on her doctoral thesis in which she studied variations of licensing processes across Europe and developed a potential structure for international licensing for SMRs.[16] Others, such as Danielle Goodman and Christian Raetzke, have suggested how models from other industries such as civil aviation could be adapted for the international certification of SMR designs.[17] They conclude their paper with a statement that I wholly support: "In a world where concerns for sufficient energy provision and GHG constraints compete with qualms about the assurance of nuclear safety, SMRs and a goal-based, risk-informed, lifelong safety regime…could support an enduring need."

9.2.3   US legal and policy barriers

The second category of institutional challenges is legal and policy barriers. For instance, Minnesota has a complete ban on nuclear power, and 12 other states in the US have statutory restrictions on the construction of new nuclear power plants: California, Connecticut, Hawaii, Illinois, Kentucky, Maine, Massachusetts, Oregon, Rhode Island, Vermont, West Virginia, and Wisconsin. Restrictions in these states vary but include requirements such as explicit approval by the state legislature, approval by the voters, or establishment of a national spent fuel repository. More subtle than these restrictions, but also challenging for utilities in regulated markets, is a ban by some states on payments for "construction work in progress" (CWIP), thus preventing the utility from recouping financing costs during the construction period. The justification for the CWIP ban is to protect the rate payer in the case of a failed construction project; however, it increases the financial risk to the utility and consequently increases the total project cost due to higher financing costs. A ban on CWIP preferentially hurts large construction projects such as nuclear plants. It will be less of an impediment to the construction of SMRs due to their lower capital cost but may still create some disincentive to consider nuclear power projects.

An issue for nuclear power that has gotten considerable attention is the electricity market distortions that are being created by federally or locally mandated policies requiring grid dispatchers to preferentially accept wind or solar-generated electricity.[18] In some instances, the amount of wind/solar electricity generation is sufficient to meet a significant fraction of the local demand, which causes base-load capacity on the grid to shut down or dump energy. The effect is to devalue base-load generating capacity, principally coal and nuclear, and to challenge the economic viability of these plants. Several utilities in the US are already experiencing increased maintenance costs for coal plants due to increased thermal cycling, and two formerly competitive nuclear plants have been permanently closed due to these market distortions.

The other major legal/policy challenge for SMRs, and all new energy technologies, is the lack of a national policy on carbon emissions. Although there has been a steady trend in the US toward restrictions on carbon and other greenhouse gas emissions, the lack of an explicit and sustainable (bipartisan) policy creates huge market uncertainty and causes energy planners to take a very short-range view on new technology investments and near-term capacity purchases.

A major step in the right direction is a set of new rules on carbon emission: the Clean Power Plan, issued by the Environmental Protection Agency in July 2015. The Clean Power Plan seeks to address both climate change and air quality impacts of carbon emissions and other air pollutants and sets aggressive goals for reducing emissions. The legislation rightfully acknowledges the value of nuclear power and will accelerate the transition to clean energy sources in the US, including nuclear energy.

9.2.4   Business challenges

Business challenges, that is, attracting sufficient funding and managing the complex endeavor of bringing a new design to market, may actually trump all other issues for the honor of highest hurdle. While I was working with the DOE to establish the federal SMR program, we asked each SMR vendor for an estimate of how much it would cost to have a new design ready for sale. They were surprisingly consistent: roughly one billion dollars. Now that I work for one of those vendors and we are substantially further down the development path, the one billion dollar figure is holding firm. Beyond the sheer magnitude of that figure are the astonishingly complex activities that consume that budget. So the first major business challenge is securing investment funds. A billion dollars is not what it used to be, but it is still a lot of money. To make matters worse, it can take more than 10 years to see a return on that investment, which is roughly how long it takes to work through all of the design development, testing, regulatory review, and detailed engineering—assuming the required funding is sustained. This places financing of new nuclear plant designs well beyond most traditional investment models.

Although some traditional large-plant vendors are developing SMR designs, presumably bankrolled by the profits from their large-plant sales and services, there have been a number of new startup companies seeking to deploy SMR designs. Many of these companies get their initial infusion of operating funds from personal investments and venture capital groups. However, this type of investment is only good for a few million dollars or maybe a few tens of millions in the case of venture capital investments. Moving beyond this startup funding is considered the "valley of death," which is survived by only a few fortunate companies. Spanning this gap generally requires moving from pure investment funds to strategic partnerships, that is, equity income in exchange for being the preferred (or exclusive) provider of a portion of the final product. In the past 10 years, more startup SMR companies have failed than have succeeded in spanning the valley of death.

Most SMR concepts that I am familiar with got their start in a research organization such as a university, laboratory, or research department within a commercial company. Only a fraction of these research concepts survive the transition into a commercial project. This creates the other major business challenge: managing a complex and protracted project that involves a huge spectrum of stakeholders, including designers, analysts, engineers, regulators, manufacturers, and suppliers. And all of this must be accomplished within a highly regimented culture of safety and quality. Growing a few-person research-based startup company to one that is capable of delivering a new certified and engineered nuclear plant design requires constant shifting of roles and responsibilities to ensure the optimum alignment of skills. Only a very few startups have accomplished this, NuScale Power being one of them.

9.3 Social challenges and opportunities

In addition to technical and institutional challenges, a number of social challenges also exist for the deployment of new SMR designs. Many social challenges, to which I also lump political challenges, are linked to the expansion of nuclear power in general, although some are specific to SMRs. First, there are too many competing SMR designs. The large number of designs and technology options creates confusion in the market with the unfortunate consequence of adding to the already high level of apprehension and caution of potential customers. We must learn from the mistakes of the first nuclear era and focus our attention on the more promising designs with an eye toward standardization. Also, designers must be honest and realistic regarding the maturity of their technology and the time line for deployment. This particular challenge may resolve itself since the high cost and multidecade process of bringing a new design to market has already caused several potential vendors to abandon or substantially curtail development efforts.

9.3.1 Nuclear waste and proliferation

Another issue that I categorize as a social challenge is the disposal of high-level nuclear waste. Although initially a technical challenge, several viable technical solutions exist, and the only remaining challenge is the social and political will to make it happen. Nuclear power is not the only industry that suffers from social and political challenges associated with waste disposal. A high-profile example is the Mobro 4000 garbage barge that in 1987 wandered the coastline from New York to Mexico to Belize and back without being allowed to dump its load.[19] The reality is that waste has a huge social stigma that creates a commensurate level of political resistance.

The US nuclear industry had a recent sobering reminder of the political vulnerability of the waste issue. After spending two decades and tens of billions of dollars to characterize the geologic environment of Yucca Mountain in Nevada and prove its suitability to store the country's commercial spent nuclear fuel, the option was abandoned virtually overnight when a different political party gained control of the US Administration and Senate in 2008. Six years later, review of the Yucca Mountain license application resumed shortly after the Senate control changed parties again in 2014. While the turnaround is encouraging, the political barrier inserted a 6-year delay in progress toward a technical solution.

The social and political challenges that ensnare the short- and long-term disposal of nuclear waste are not specific to SMRs. Furthermore, SMRs do not in themselves change the waste dilemma in either direction, that is, waste is size neutral. On the other hand, different reactor technologies can change the volume and characteristics of the discharged waste. For example, metal-cooled reactors can be designed to partially consume or transmute hazardous waste during normal operation. Since SMR designs span the same reactor technologies as large-plant designs, they produce similar types of waste at roughly the same rate as large reactors of the same technology when normalized to the same total power. Because of the fact that the nuclear waste issue is size neutral, I will not address the topic further, except to acknowledge that it is a huge social and political challenge for the nuclear industry. A significant step forward in addressing this challenge was a recommendation by President Obama's appointed Blue Ribbon Commission that the US should move to a consensus-based approach for the siting of interim waste storage facilities and a final repository.[20] If the actual risks of hosting a storage or repository facility are appropriately communicated to the local constituency and suitable incentives are included, I believe the long-standing challenge of siting a repository will finally be resolved.

Another social/political challenge that SMRs share with their bigger siblings is the concern for the proliferation of nuclear weapons. Nuclear proliferation is a complex, controversial, and for some a very emotional issue. Every study that has been conducted on nuclear proliferation has come to the same conclusion: reducing the proliferation risk requires both technical and institutional fixes, with an emphasis on institutional. I cannot reasonably do the topic justice and will not attempt to do so, except to offer a couple of high-level observations regarding the impact of SMRs on the issue. Some believe that commercial nuclear power is a logical pathway for nuclear proliferation. I do not share this opinion, but for those who do, SMRs may exacerbate the concern since the affordability of SMRs can lower the threshold for new countries to initiate commercial nuclear power, resulting in more plants in more places. Balancing this fact is the opportunity to engage with these countries from the outset to influence how their fledgling nuclear program is implemented and ensure that the technology is used correctly. The consequences of nuclear proliferation can be disastrous, but so can the consequences of not meeting the rapidly growing global need for energy. We must address both potential consequences with great care.

9.3.2   Surviving opponents and supporters

SMRs are not without their opponents. To date, however, most anti-SMR arguments have come from long-standing antinuclear individuals and groups, so their arguments are not surprising, and their impacts have been largely inconsequential. A former colleague shared some wisdom with me many years ago: "It's easy to throw stones at an idea; rocks are cheap." Most opponents have nothing to offer except criticism. I am actually surprised and pleased at the rather benign nature of the criticisms that have been levied at SMRs. Most of the complaints are that their advertised benefits have not been demonstrated yet, which is a valid observation but of little value—we already know that. Some opponents challenge the unproven safety benefits while others assert that SMRs cannot be economically competitive, typically without offering any supporting evidence. On the positive side, anti-SMR literature can have some value. First, it serves to keep the dialogue on SMRs open and transparent. It may also offer us some insight into legitimate issues with the technology that require additional study or better articulation. At the same time, it is important to understand the underlying motivations of the various naysayers in order to properly interpret the relevance and significance of their objections.

Related to the opponents are those that feel threatened or disenfranchised by the growing interest in SMRs. It is reasonable to expect some level of backlash from those promoting the incumbent technologies or those developing alternative products. I have especially noticed this reaction by those who are entrenched in the large-reactor business. To be honest, some SMR advocates have unintentionally encouraged this backlash reaction. Too often we carelessly use broad assertions like "safer than" and "cheaper than" inappropriately when making comparisons to large plants. The reality is that SMRs and large reactors can be quite complementary and support different customers. Those customers generally require larger safety margins and more affordability, which makes SMRs very attractive. There have also been instances of anti-SMR backlash from non-nuclear energy sectors that appear to resent the high level of interest and federal funding going toward SMRs rather than their favorite technologies, typically wind or solar. This is another example of inappropriate comparisons that treat SMRs and renewables as though they are interchangeable, which they are not.

Of the many different types of individuals that challenge the value proposition of SMRs, I enjoy interacting most with the skeptics. Although sometimes lumped with opponents, they are quite different. Skeptics ask pointed questions and listen openly to the answers. They quickly see past "smoke screens" and focus on the fundamental issues. Most importantly, they are willing to change their opinions if properly convinced, and if not, will share the basis for the tenacity of their opinions. My personal experience has been that a high percentage of SMR skeptics become advocates when properly engaged and do so with a strong basis of understanding.

It might seem odd to include "supporters" in a chapter on challenges, especially grouped under the same heading as opponents. For sure, SMRs have broad and growing support across the industry, and this support is critical for moving SMRs forward. However, there are sometimes unintended consequences from that support, especially in the case of SMR zealots. In this context, I define a zealot as someone who has only one answer—his or her answer—regardless of the question. There are a number of reasons why I highlight this group under social challenges. First, they can be deceptively misleading. They put forth a very positive presentation in favor of SMRs in general or for a specific design and can articulate the benefits exceedingly well. Unfortunately, they typically do not see, or are unwilling to admit, any shortcomings. They present their specific solution as the "silver bullet" that everyone hopes for but rarely exists. A second concern is that zealots tend to be highly competitive and frequently discredit other solutions unfairly, which leads to very confusing messages for the broader community. My third and greatest concern for over-the-top enthusiasts is the hyperbole that creates false expectations about what SMRs can do and the schedule on which they can do it. An unrealistic optimism that ignores the technological, regulatory, and commercial challenges of bringing a new technology to market serves only to confuse potential customers and dilute the industry's focus. It also generates disillusionment and encourages the labeling of SMRs as a short-term fad that cannot deliver on promises.

9.3.3 Public acceptance

Although it is common in the nuclear industry to include the topic of public acceptance in the category of challenges, I present it as an opportunity. The reality is that public approval of nuclear power in the US has increased steadily over the past four decades and is now solidly favorable. But those of us who have worked in the field for many years remember a time when it was best to keep our profession to ourselves in casual gatherings. So there is a lingering tendency to perpetuate the notion that nuclear power has a low public acceptance rating. In my review of the literature on public acceptance of nuclear power, both the pronuclear and the antinuclear organizations consistently observe an upward trend, although

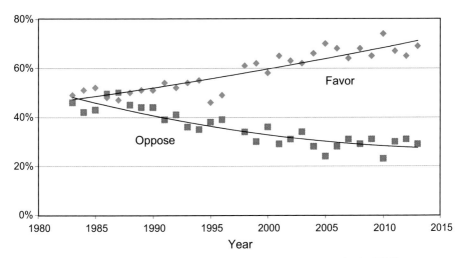

Figure 9.1 Support by the general public for the use of nuclear power in the US.[21]

the absolute ratings differ between the two sources. I tend to favor the data from organizations such as the NEI, not only because their results are more encouraging but also because they are more forthcoming with details on how the data were acquired and interpreted.

Figure 9.1 provides the results of a sustained polling of the general public conducted by Bisconti Research on behalf of the NEI.[21] Since the sample set for each poll was typically 1000 people, one would expect a 3–6% (one to two standard deviations) scatter in the data. I have added polynomial trend lines to help smooth out the data scatter. What is especially encouraging about the results is that the nominal 75% approval rating for the past few years is quite uniform among respondents, that is, independent of gender, age, or political party. Based on these consistent findings, we in the industry need to get over the old stigma and graciously accept the new reality that nuclear power is now accepted by a clear majority of the general public in the US. Our challenge is to stay vigilant about maintaining the excellent safety and performance record of the operational fleet and ensure that new plants are at least equally successful.

It is interesting to note that a survey question regarding the perceived level of safety of operating nuclear plants produced data that are nearly identical to the curves shown in Figure 9.1. I suspect that this is not a coincidence—the general public appears to be well aware of the safety record of US nuclear plants. Even in 2011, the year that the Fukushima Daiichi nuclear plant in Japan was destroyed by an earthquake and tsunami, support for nuclear power in the US only dropped from 71% at the beginning of the year to 65% at the end of the year and completely rebounded back to 70% by 2013. My assessment of this is that the general public was perceptive enough to realize that the disaster in Japan was the horrific tsunami, not the damaged Fukushima plants, despite attempts by some to portray it differently.

Of real significance is a growing movement among environmentalists to recognize the merits of nuclear power in combating environmental concerns. Patrick Moore, cofounder of Greenpeace, was one of the first major environmentalists to speak out in favor of nuclear power. In 2013, a Hollywood-quality documentary film was released—*Pandora's Promise*, directed by Robert Stone—that discusses the concerns and promises of nuclear energy. The film includes several notable environmentalists who previously rejected nuclear power but now acknowledge it as an essential component in the fight against global warming. One of the environmentalists included in the film, Gwyneth Cravens, published a book in 2008 entitled *The Power to Save the World: The Truth About Nuclear Energy*.[22] What I found especially interesting about her book was that her conversion from a nuclear opponent to supporter occurred only after a thorough investigation of the technology from mining of the uranium ore to disposal of the nuclear waste. This should give the nuclear industry a valuable clue on the merits of doing a better job with nuclear education.

There is much more that the industry can do to further improve and solidify public support for nuclear power, especially in the area of communication and education. For instance, in the most recent survey, only 13% of the participants indicated that they were "well informed" about nuclear energy. Those who live within 10 miles of a nuclear plant indicated a higher level of knowledge about nuclear energy and also had a higher approval rating for the use of nuclear power than the general public. As an example, an impressive 91% of nuclear plant neighbors supported the renewal of operating licenses for near-by plants.[23]

I believe that we have an opportunity with SMRs to push public acceptance of nuclear power to unprecedented levels. But just as SMRs bring new innovation to nuclear plant designs, we also need to introduce new approaches to how we communicate nuclear energy to the general public. Engineers are dismal at communication, so we will need to engage social scientists and communication experts to serve as a liaison with the public and translate our jargon. An excellent study by the Nuclear Energy Agency in 2002 concluded that the average citizen perceives risk very different than the expert.[24] While we in the industry communicate risk in terms of mathematical equations and probability distributions, the general public's perception of risk is a complex balancing of more than a dozen qualitative factors, including trust, control, voluntary versus involuntary, benefit/reward, understanding, and even gender. With the help of social scientists and communications specialists as well as direct engagement of the general public, I believe that we can succeed in conveying the message that SMRs offer clean, abundant power in an astonishingly robust package. I am especially optimistic about the opportunity for SMRs to change how the public thinks of nuclear power in places that do not currently have nuclear plants. In those places, we have the opportunity to get it right the first time, instead of spending the next 40 years recovering from a bad start.

9.4 Government roles

The time-honored cliché of "we're from the government and we're here to help" never fails to bring a snicker in any crowd. Yet within the nuclear industry, the government has and should play a substantial role. The US is almost singularly unique worldwide in the degree of separation between the government and the nuclear industry. In most other countries, the government, the nuclear R&D laboratories, the nuclear vendors, and the utilities are one and the same. A cynic would say that this puts the US at an advantage, and I must admit feeling this way at times. But the reality is that the separation of state and industry is a tremendous disadvantage, especially with respect to global competitiveness.

Although the US government is not intimately linked to the commercial nuclear industry, it has played pivotal roles in initiating and sustaining nuclear power development. Likewise, the government has been a sustained supporter of SMR development and deployment throughout the history of nuclear power in the US. I discussed in Chapters 2 and 3 some of the early roles that the government had in promoting smaller nuclear plant designs in the 1980s and 1990s, including the Advanced Light-Water Reactor and the Advanced Liquid-Metal Reactor programs. In the early 2000s, the Nuclear Energy Research Initiative spawned several SMR designs, and the Generation IV program included peripheral support of SMRs. Beginning in 2006, the DOE introduced a series of programs specifically targeting SMRs, including the Grid-Appropriate Reactors program, which was an element of the Global Nuclear Energy Partnership, the SMR Licensing Technical Support (LTS) program, and the SMR R&D program. Table 9.2 provides a summary of key DOE programs that have contributed to the development and eventual deployment of SMRs in the US.

The last two programs listed in Table 9.2 are of special importance in terms of meeting the challenges presented earlier in this chapter. The SMR LTS program was first funded in 2012 for the purpose of accelerating the design completion, licensing, and deployment processes of near-term SMR designs that have the potential to achieve commercial operation by roughly 2021–2025. Funds were awarded to two separate SMR vendors: Generation mPower in 2012 and NuScale Power in 2013. The 5-year program shares the costs of obtaining design certification approval from the NRC and also the completion of first-of-a-kind design and engineering. The program expects to resolve many of the technical and licensing challenges discussed earlier in this chapter.

Also initiated in 2012 was the SMR R&D program, which was merged into the Advanced Reactor Concepts program in 2015. The purpose of this program is to support applied research for the development of non-LWR SMRs, such as gas, sodium, lead, and salt-cooled SMRs. In addition to R&D on various cross-cutting technologies, the Advanced Reactor Concepts program is also coordinating an effort with the NRC to develop technology-neutral general design criteria.[25] This effort, which builds upon the SMR licensing demonstrations in the LTS program, will facilitate the eventual licensing of designs based on the advanced technologies.

In addition to supporting relevant research and licensing activities, the federal government can substantially influence the likelihood of success in deploying SMRs, or nuclear power in general, through various legal and policy implementations. A significant boost for nuclear power was the Energy Policy Act (EPAC) that became law in 2005.[26] The EPAC extended or introduced several incentives for nuclear power,

Table 9.2 **DOE programs supporting research, development, and deployment of SMRs**

Program	Period	Relevance to SMRs
Advanced Light-Water Reactors	1982–1997	Encouraged the development of smaller plant designs and the use of passive safety systems. Resulted in the certification of AP-600 and ABWR designs.
Advanced Liquid-Metal Reactors	1984–1994	Encouraged a new look at sodium-cooled reactors. Resulted in the development of the PRISM SMR design.
Nuclear Energy Research Initiative	1999–2007	Funded several SMR project teams, including LWR and non-LWR designs.
Generation IV	2000–pres.	Created a framework for advanced reactor research and re-energized the US R&D community.
Next Generation Nuclear Plant	2005–2012	Pursued the development of a small, high-temperature reactor for process heat applications.
Nuclear Power 2010	2002–2010	Demonstrated the new 10 CFR, Part 52 licensing process and re-energized the US nuclear industry.
Nuclear Energy Universities Projects	2009–pres.	Funds university-led projects, including R&D relevant to SMRs.
GNEP/Grid-Appropriate Reactors	2006–2008	Stimulated the design of SMRs focused on export to developing/emerging countries.
SMR LTS	2012–pres.	Accelerating the licensing process and first-of-a-kind engineering for near-term SMRs.
SMR R&D/Advanced Reactor Concepts	2012–pres.	Funds applied research for the development of advanced (non-LWR) SMR technologies and designs.

including research, investment, and production incentives. Primary provisions of the EPAC include the following:

- continued authorization of the Nuclear Power 2010 program to encourage design completion and licensing of new nuclear plants,
- loan guarantees to reduce the cost of borrowed money to finance new nuclear projects,
- delay insurance to protect new plant owners from increased project costs due to licensing delays,
- production tax credits for the initial period of operation of new nuclear plants,
- limited liability in the case of an accident at a nuclear plant, and
- reduced tax rate for decommissioning funds.

A study by the Congressional Business Office in 2008 evaluated the value of the EPAC 2005 legislation in helping to encourage investments in new nuclear plants and showed that the various provisions substantially helped to "level the playing field" with other base-load energy options.[27] However, the specific provisions implicitly targeted the deployment of large Generation III nuclear plant designs such as the AP-1000 and the ESBWR. Time limitations specified for some of the incentives preclude them from applying to SMRs. For example, to benefit from the production tax credit, the first pour of safety-grade concrete had to be performed by 2014, and the plant must be operational by 2021, which appears to be unlikely for SMRs that are still in the design phase. Therefore, an update to EPAC 2005 is needed with provisions and time limits more aligned with the deployment of near-term SMRs. This was partially addressed in a draft solicitation issued by the DOE Loan Program Office in October 2014, which provides up to $12.6 billion in loan guarantees for innovative nuclear energy projects, including SMRs.

An important role and significant challenge for the US government is to develop a sustainable policy on carbon and other air pollutants emitted across all sectors, including electricity production, transportation, and industry. As mentioned earlier, the Environmental Protection Agency has issued a set of new rules to curtail emissions, but the staying power of these rules and the government's conviction to enforce them remains to be proven. Still, it is clearly the role of the government to take the long vision and implement policies that can get us there. It is essential that they succeed.

Finally, the government has the need and opportunity to become a first-mover for the deployment of SMRs to meet the energy demands of government facilities while significantly improving the clean energy mix of these facilities. The government is the single largest user of energy in the country and has an opportunity to lead the way in the deployment of clean, abundant energy from SMRs. The DOE and the Department of Defense have taken a serious look at doing this, and the DOE is moving forward with plans to do so. In particular, the DOE is engaged with SMR vendors and utilities to potentially site an SMR adjacent to the Oak Ridge National Laboratory and also on or near the federal reservation of the Idaho National Laboratory. Both of these laboratories have had many research and test reactors located at their sites in the past, and each has a research reactor currently operating there. However, unlike the research reactors that are owned and regulated by the DOE, the SMRs will be owned by a utility and regulated by the NRC as commercial nuclear plants. Long-term power purchase agreements with the DOE will help to offset the first-of-a-kind risk premiums that the utilities will pay for the new designs. In turn, these first-of-a-kind SMR plants will help to reduce the anxiety of potential follow-on customers, both federal and commercial.

There are many remaining challenges for the successful deployment of SMRs, especially in the US. However, these challenges bring opportunities for improvement and innovation in technology, regulation, and social acceptance. I am encouraged by the level of excitement in the nuclear industry and the breadth of public–private cooperation. When my friends in the industry ask how they can help to get SMRs to the finish line, my stock answer has become "patience and generosity." For sure, it is a long and expensive path, but the potential benefits to the US and the world are enormous.

References

1. Thomas S. *The EPR in crisis*. PSIRU Business School, University of Greenwich; November 2010.
2. Ingersoll DT, Poore III WP. *Reactor technology options for near-term deployment of GNEP grid-appropriate reactors*. Oak Ridge National Laboratory; 2007. ORNL/TM-2007/157.
3. Carelli MD, Ingersoll DT. *Handbook of small modular nuclear reactors*. Cambridge, UK: Woodhead Publishing; 2014.
4. Houser R, Young E, Rasmussen A. Overview of NuScale testing program. *Trans Am Nucl Soc* 2013;**109**:1585–6.
5. *CAREM construction underway*. World Nuclear News; February 10, 2014. Available at: www.world-nuclear-news.org/NN-Construction-of-CAREM-underway-1002144.html.
6. *Potential policy, licensing and key technical issues for small modular reactors*. US Nuclear Regulatory Commission; March 28, 2010. SECY-10–0034.
7. *Proposed improvements to Tier 1 and the inspections, tests, analyses, and acceptance criteria (ITAAC) for small modular reactors*. Nuclear Energy Institute; March 14, 2014.
8. *Development of an emergency planning and preparedness framework for small modular reactors*. US Nuclear Regulatory Commission; October 28, 2011. SECY-11–0152.
9. *Proposed methodology and criteria for establishing the technical basis for small modular reactor emergency planning zone*. Nuclear Energy Institute; December 23, 2013.
10. *Use of risk insights to enhance the safety focus of small modular reactor reviews*. US Nuclear Regulatory Commission; February 18, 2011. SECY-11–0024.
11. *Guidance for assessing exemption requests from the nuclear power plant licensed operator staffing requirements specified in 10 CFR 50.54(m)*. US Nuclear Regulatory Commission, NUREG-1791; 2005.
12. *Status of the office of new reactors readiness to review small modular reactor applications*. US Nuclear Regulatory Commission; August 28, 2014. SECY-14–0095.
13. *Multinational design evaluation program annual report: March 2013–March 2014*. Nuclear Energy Agency; April 2014.
14. *Facilitating international licensing of small modular reactors*. Cooperation in Reactor Design Evaluation and Licensing Working Group; 2015.
15. *"Proceedings of the 6th INPRO dialog forum," International Atomic Energy Agency, July 29–August 2, 2013*.
16. Söderholm K, Amaba B, Lestinen V. Licensing process development for SMRs: European perspective. *Proceedings of the ASME 2014 small modular reactors symposium, Washington, D.C., April 15–17, 2014*.
17. Goodman D, Raetzke C. SMRs: the vehicle for an international licensing framework? a possible model. *Nucl Future* 2013;**9**(6).
18. Banks GD. *The unintended consequences of energy mandates and subsidies on America's civil nuclear fleet*. Center for Strategic and International Studies; May 13, 2013.
19. Katz J. What a waste. Reg Rev, Q1 2002.
20. *Report to the Secretary of Energy*. The Blue Ribbon Commission on America's Nuclear Future; January 2012.
21. *Perspective on public opinion*, prepared by Bisconti Research for the Nuclear Energy Institute; October 2013.
22. Cravens. *Power to save the world: the truth about nuclear energy*. New York: Alfred A. Knopf; 2008.
23. *Favorability toward nuclear energy stronger among plant neighbors than general public*. Bisconti Research; Summer 2013.

24. *Society and nuclear energy: towards a better understanding*. Nuclear Energy Agency; 2002.

25. *Guidance for developing principal design criteria for advanced (non-light water) reactors*. Idaho National Laboratory; 2014. INL/EXT-14–31179, Rev.1.

26. *Energy policy act of 2005*. United States Government; August 8, 2005. Public Law 109–58.

27. *Nuclear Power's role in generating electricity*. Congressional Budget Office; May 2008.

Fad or future?

I began this book freely admitting my bias toward nuclear power. I reiterate that bias here: I am a huge fan of nuclear power and what it has to offer the world in terms of clean, abundant energy. I am also a huge fan of small modular reactors (SMRs) as a way to enable nuclear energy to fully achieve its promise. In the early chapters, we saw a sustained interest in SMRs from the very inception of the nuclear industry, which should be sufficiently convincing that they are not just a fad. In subsequent chapters, I described many compelling benefits of SMRs, including enhanced safety and robustness, improved affordability, and increased flexibility. I even provided evidence of substantial interest by a diverse group of potential customers. But an important question remains: Can they finally become a commercial reality and more importantly, will they become a significant part of nuclear power's future?

10.1 The fad

There are legitimate reasons for thinking that the current burst of interest in SMRs might represent a fad. Despite persistent interest in SMRs throughout the 60+ year history of commercial nuclear power, they have no significant presence in the existing global fleet of nearly 450 nuclear power plants. Previous chapters speak to why this might have happened, but the fact remains that SMRs are not a part of nuclear power's present. Then why should we be optimistic that they will be a part of its future? Even fads have their moment of achievement, albeit short. SMRs cannot yet boast even this accomplishment.

I originally thought that the formative years of SMR's latest incarnation, that is, roughly the period from 2000 through 2009, might be the most vulnerable time for SMRs since they were largely unknown and in very early stages of development. I now believe that the past 5 years have presented a higher vulnerability due to a combination of two very different factors: high expectations for the success of SMRs and increasing doubt regarding the economic viability of nuclear energy in general. The latter vulnerability results from a variety of economic and political circumstances that make it difficult for even well-performing base-load plants to be competitive in markets that have abundant, cheap natural gas and policies that favor wind and solar power generation. These issues have been discussed in previous chapters, so I will not relive those challenges here.

The vulnerability for SMRs created by high expectations can best be described using the "hype cycle," a construct developed by business advisors at Gartner, Inc. in 1995.[1] According to the hype cycle, a new product or technology experiences

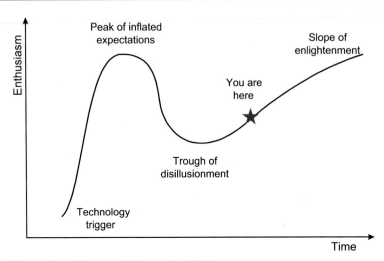

Figure 10.1 Notional "hype cycle" curve for SMRs.

an initial rapid growth of interest and excitement, which then peaks and falls as overly optimistic expectations are unmet. Eventually the disillusionment bottoms out. For those technology providers that make it through the "trough of disillusionment," interest and excitement in the technology again increases as progress is realized. I have depicted in Figure 10.1 the hype cycle for SMRs, including the proverbial "you are here" marker to indicate my estimate of where the US nuclear industry is currently within the cycle.

I associate the announcement of the Department of Energy's SMR program in 2009 as the "technology trigger" that started the existing SMR hype cycle. It was shortly after that announcement that several new SMR designs emerged from various organizations. At the same time, the nuclear industry became alive with conferences and media coverage on the subject. Established technical societies such as the American Nuclear Society began sponsoring sessions and then whole conferences devoted to SMRs, and several for-profit conference organizers jumped on the bandwagon to host high-profile conferences targeting SMRs. Unfortunately, the hyperbole that accompanied those conferences created an expectation that SMRs were going to be the savior of the nuclear renaissance and that they were zipping along toward imminent license approval and construction.

By 2013, the conference circuit had become saturated with SMR events—many of which included the same speakers presenting basically the same content with only minor incremental progress. That is when reality set in, and the community began to recognize that bringing a new design to market, even a smaller, simpler design, was a very prolonged and costly endeavor. Then in late 2013 and early 2014, Westinghouse and Babcock & Wilcox substantially reduced their level of effort in developing their SMR designs, citing uncertain market opportunities or investment challenges. I believe that these events triggered the decline from the "peak of inflated expectations." It was during the several months following these announcements that the number of SMR-related stories posted on the Internet dropped precipitously. Also, the tone of several media articles turned more pessimistic and foreboding regarding the eventual outcome for SMRs.

However, there is solid evidence that the disillusionment part of the cycle was short-lived and that the SMR industry is already progressing into the "slope of enlightenment." Several things happened in the latter half of 2014 and early 2015 to quickly restore interest in SMRs, although the resumed interest appears to come with a more mature and realistic expectation. Some of the key confidence-building events include the following:

- The Tennessee Valley Authority announced that they were continuing to proceed with efforts to obtain an early site permit from the US Nuclear Regulatory Commission (NRC) for the Clinch River site despite the fact that Generation mPower had substantially curtailed their design and licensing activities.
- A Washington state senator introduced several bills to the state legislature to pave the way for deploying SMRs in that state.
- The Utah Associated Municipal Power Systems consortium announced that they were proceeding with their Carbon-Free Power Project, which includes the deployment of a NuScale SMR plant in Idaho.
- Westinghouse announced the approval by the NRC of a topical report that they had submitted previously, implying that they were still active in pursuing their SMR design.
- President Obama issued an Executive Order directing federal facilities to dramatically reduce their carbon emissions, explicitly citing SMRs as a clean energy alternative.

In addition to these US events, interest in SMRs continued to show progress worldwide. In particular, Argentina began construction of their CAREM prototype SMR, China resumed construction of their HTR-PM demonstration project, and the United Kingdom completed a feasibility assessment for SMRs motivated by the "recognition from Government of a need for further industrial development, and for low carbon, secure and affordable energy supply."[2] The combined impact of these events, domestic and globally, gives me confidence that SMRs have the staying power to succeed.

10.2   The future

So if SMRs are not a fad, do they have a future? I believe that they do and that they will become a growing part of the global nuclear power fleet. I base this assertion on a number of considerations. First and foremost, I have to believe that nuclear power will continue to be a major contributor to the domestic and global energy mix. The growing demand for energy and the increasing concerns for carbon emissions will drive the US and most countries to expand clean energy options. Nuclear power currently supplies over two-thirds of the US' carbon-free energy and is the only emissions-free source that can provide the magnitude and reliability of energy that will be required to run our industrial infrastructure. One need only look at the unfolding economic challenges in Japan and Germany resulting from their recent curtailment of nuclear power to understand its potential impacts. In these two countries, electricity prices have soared, carbon emissions have risen, and energy imports have increased—not a sustainable strategy.[3] Hence, nuclear energy is with us to stay.

Within the context of a growing nuclear power future, SMRs fill an important global need: they offer clean, abundant, and reliable energy in a package that is affordable, flexible, and highly robust. As discussed in Chapter 8, SMRs are not a technology push but rather a response to compelling customer needs and requirements—needs that cannot be met effectively with large nuclear plants or alternative energy sources. When I survey the domestic and international interest in SMRs, it comes from small utilities, large utilities, utilities that are new to the nuclear power community, and utilities that are nuclear veterans. The basis for their interest may be different, but they uniformly understand the potential benefits that SMRs have to offer. Perhaps these broad and diverse interests are why the "trough of disillusionment" in the SMR hype cycle was so brief and shallow.

Another aspect of SMRs that gives me optimism is that there are many successful analogs in other industries. I discussed earlier the revolution in the computing industry during the early 1990s that transitioned the computing community from huge (and expensive) single-processor mainframes to the massively parallel supercomputers of today. The airline industry also experienced a shift from the ever-larger jumbo jets of the 1990s to the pervasive use of regional jets during the past decade. Henry Ford brought automobiles to the masses 100 years ago by introducing a new approach to manufacturing: assembly line production. In the energy industry, coal plants shifted to modular plants in the 1970s and 1980s to overcome an increasing maintenance problem with the progressively larger units. It is time to bring modularity to the nuclear industry.

I fully support the continued operation and new construction of large nuclear plants. They serve the needs of those customers who can afford them. Also, expanding the use of nuclear energy using SMRs would not be an option today were it not for the high safety and performance record of the existing large-plant fleet. But I also share a growing concern with many others that the future of new large plants is in jeopardy. We are seeing trends today reminiscent of the original buildout in the 1960s and 1970s—trends in construction delays and cost overruns. I worry about these trends and what they mean for the future of nuclear energy. This drives me to believe that it is time to break the mold and deploy nuclear power in a very different way. I believe that SMRs offer that opportunity to manufacture and deploy nuclear power with restored certainty and confidence. I also believe that SMRs, that is, small, simple, and robust packages, are the way that nuclear energy is meant to be used. Finally, I believe that SMRs will be an important part of our clean energy future. A lot of talented people are working hard to make it happen, and I am honored to be working among them.

Still, there are many remaining activities and challenges for even those SMR designs that are furthest along their deployment path. The designs need to be completed and certified by the NRC. Manufacturing facilities need to be tooled and the supply chain finalized. Most importantly, the first courageous customer needs to step forward and write the first check. None of these steps are trivial or cheap. So we are a few years, at best, from knowing for sure if SMRs have a future. I remain confident that they do.

10.3   Looking beyond the future

Albert Einstein has been quoted as saying the following: "Logic will get you from A to B. Imagination will take you everywhere." In the first 60 years of commercial nuclear power, logic has gotten us to the present situation of a highly successful global fleet of large power plants. Imagination, the kind of creativity that is being used by SMR designers worldwide, will take us to a more expansive future for nuclear energy. In this section, I offer a look beyond the future, that is, beyond the first successful operation of an SMR. Assuming that the first SMRs succeed at getting built and that their promises are realized, what could their future look like beyond that?

From the perspectives of technology and regulatory readiness, SMR designs based on traditional water-cooled reactor technology will likely be the first to achieve commercial deployment. Blazing that trail will open the door for future SMR designs with more innovative features and alternative coolants. As discussed in Chapter 4, reactor designs using gas or salt coolants can operate at much higher temperatures than water-cooled reactors, which may be a better solution for some process heat applications. Also, metal coolants such as sodium and lead allow the reactor to run on "fast neutrons" and enable applications such as fuel breeding and nuclear waste consumption. Both high-temperature and fuel-cycle applications can also be served by large reactor designs; however, packaging them as SMRs can add safety, affordability, and flexibility to their use as it does with small water-cooled reactors. Even for water-cooled SMRs, new fuels, materials, and components can further increase their safety and performance.

These advanced SMRs will further expand the market opportunities for nuclear power. For instance, metal-cooled reactors can be designed with very long fuel lifetimes, perhaps even 30 years. They do this by producing fuel within the core at a rate comparable to the consumption of the initial fuel loading. This will be of significant interest to communities or installations that are difficult to reach or that require long-term continuous power. Also, high-temperature SMRs will have higher power conversion efficiencies and therefore dump less heat to the environment, thus reducing water withdrawal and consumption—important considerations in very arid regions.

Although the first SMRs to be deployed are likely to be for electricity generation in traditional markets, their robustness and flexibilities will be very attractive to nonelectrical energy consumers, that is, process heat customers. I anticipate rapidly escalating interest in SMRs for district heating and cooling, water desalination, and a variety of industrial heat applications. This will happen first using cogeneration operations; that is, the SMR will be directly coupled to a single industrial plant and provide both electricity and steam to the plant. Eventually, we may see complex systems of heat and electricity generators (nuclear, solar, wind, hydro, etc.) working together to supply electricity and heat to a network of energy consumers. These hybrid energy systems will enable a high level of system optimization to minimize resource consumption and maximize product value. In a Forbes column, Jim Conca asserted a similar vision:

> *Whether it's SMR's ability to load-follow renewables, desalinate seawater, provide district heating, power chemical production, oil refining, hydrogen production or advanced steelmaking without the emissions presently released by fossil fuel plants, co-generation and hybrid systems are key ingredients to our global energy future.*[4]

After the first few SMRs are constructed, probably using existing manufacturing capacity, there will come a time when a new factory is justified. This will open up a new spectrum of opportunities associated with advanced manufacturing technologies enabled by the fact that the modules are built in a factory. Advanced joining and cladding technologies and maybe even three-dimensional printing methods will enable the modules to be built better, faster, and cheaper.

Similarly, the higher degree of standardization that a modular plant allows can enable highly streamlined certification of new modules. This standardization will also facilitate more uniform operator training and perhaps allow fleet-wide operator certification, including international fleets. With this standardized approach will surely come global fleet management strategies similar to what is done by manufacturers of jet engines for a global fleet of commercial airliners. To be sure, these ideas are lofty concepts that are decades away from realization, but we have to start sometime. I suggest now.

I finish my story where I began: the need for energy. The people of the world have many diverse circumstances and energy needs. While we worry about where to charge our cell phones, roughly one-third of the world's population does not have access to electricity. More than a billion people live in extreme poverty, and more than two billion people lack access to sanitary water.[5] We can predict several future outcomes with a high level of certainty: the global population will continue to grow, everyone will strive for a higher quality of life, and water and energy resources will become increasingly scarce while their demand will increase. The energy intensity inside the nucleus is staggering, yet we have learned how to release it and to control it. We have also learned how to put it to work but have only scratched the surface of its full potential. SMRs are the next step in expanding the utility of nuclear energy to more people in more places to meet their fundamental needs of energy and water. SMRs have the opportunity and the potential to improve the world, and now is the time to make it happen.

References

1. The Gartner hype cycle, Gartner, Inc., Available at: www.gartner.com/technology/research/methodologies/hype-cycle.jsp [accessed 26.04.15.].
2. *Small modular reactors (SMR) feasibility study.* UK National Nuclear Laboratory; December 2014.
3. Alexander L. The United States without nuclear power, speech delivered to the Nuclear Energy Institute. February 5, 2015.
4. Conca J. *Can SMRs Lead the U.S. into a clean energy future?.* Forbes; February 16, 2015. Available at: www.forbes.com/sites/jamesconca/2015/02/16/can-smrs-lead-the-u-s-into-a-clean-energy-future/.
5. *Water for a sustainable world,* United Nations world water development report. 2015.

Index

Note: Page numbers followed by "f" and "t" indicate figures and tables, respectively.

Printed in the United States
By Bookmasters